ASE Secondary Science Teachers' Handbook

Edited by Richard Hull

SIMON & SCHUSTER
EDUCATION

Acknowledgements

The following people have contributed to this handbook: Rosemary Sherrington, editor the companion primary handbook, which shares some chapters with this one: John Alle John Holman, Bob Male, Sheila Martin, and James Williams who read and commented sections and particularly Bryan Milner who read and commented on most of the boo David Bevan, then Chair of ASE Publications Committee, who gave advice: both him ar Dave Roddis who gave technical help in changing texts on discs into a form acceptable the editor's computer: Valerie Keye who re-keyed a number of chapters that defeated eve them, Alan Jones who helped with Chapter 10: and Saxon Hull who helped with the who book.

Kind permission has been granted for the use of the following extracts in this book:
page 48 Figure 1 was first published in *Interactive teaching in science: workshops f training courses 9. Diagnostic teaching in science classrooms* and is reproduced by permi sion of the Controller of Her Majesty's Stationery Office.
page 67 Figure 1 was first published in NCC INSET Resources *'Science Exploration* and is reproduced by permission of the National Curriculum Council.
page 82 Table 1 DES (1992) 'Reporting pupils' achievements to parents', Circular 5/92
page 265 Figure 1 First published by Worthing High School.
pages 322 and 323 The material in Tables 5 and 6 was taken and adapted from *No statutory Guidance for Geography* and is reproduced by permission of the Nation Curriculum Council.
The ASE has tried to contact all other copyright holders and would be happy to hear fro anyone whose rights they have unwittingly infringed.

The views expressed in this book are those of the authors and not necessarily those of the employers.

© The Association for Science Education, 1993

First published in Great Britain in 1993 by
Simon and Schuster Education
Campus 400, Maylands Avenue
Hemel Hempstead, Herts HP2 7EZ

Reprinted in 1993, 1994

Printed in Great Britain by
Redwood Books, Trowbridge, Wiltshire

A catalogue record of this book is
available from the British Library

ISBN 0 7501 0449 X

Contents

Introduction

Much has changed in science education since the publication of the first ASE Science Teachers' Handbook in 1986. Not least is the growth of Primary Science which has necessitated the production of the ASE Primary Science Teachers' Handbook in parallel with this one.

The introduction to the old handbook said that "Science teachers are busy people caught up in a rapidly changing educational scene" If we were then, you should see us now! Think what it was like.

No LMS, no NCC,
No SEAC or the DFE,
.....

(An appropriate tune and subsequent lines may occur to the reader!)

The intention of the present handbook is to give an impression of where we are now in Science Education, with some reflection on how we got here, for the pace has been hectic and there has been little time for reflection. Readers will no doubt vary widely in their interests, so the book looks at Science Education with a varying breadth of vision. The idea is illustrated in the diagram.

The closest focus is on the classroom and laboratory, in Chapters 3, 4 and 5, which look at teaching, learning, assessment and evaluation. It is perhaps here, initially, that a teacher in training or a new teacher might want to start.

We then zoom out a little and focus on the Science Department. In the "Managing and Organising in Science Education" chapters, we are concerned with management and organisation generally (Chapter 6) and then look particularly at safety, and the use of new technology and of living organisms (Chapters 7, 8 and 9). In the "Science for All Pupils" chapters, the concern is equal opportunities; for all races and both sexes, and for pupils with special needs.(Chapters 10 to 12).

The third circle indicates a broadening out to see secondary science education in the context of "Science for All Ages", from the nursery right through to age 18. Here the authors are giving a flavour of what science is like at the various levels and, additionally, there is a focus on continuity and progression (Chapters 13 to 16).

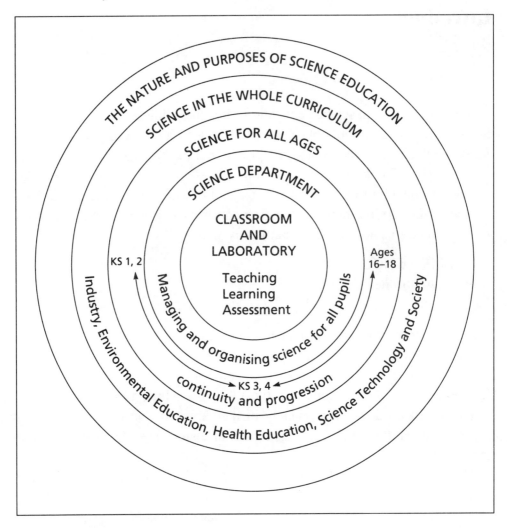

Focuses in Science Education

In the fourth circle, the focus is the curriculum, the Science National Curriculum in Chapter 2, and science and its connection with environmental and health education, with technology and society, and with economic and industrial understanding, in Chapters 17 to 20.

Finally, with the widest view of all, Chapters 1 and 21, bracketing the other chapters in the lay-out of the book, examine what the whole enterprise of science education is about.

There are various themes that run through the book that we might envisage as spokes joining the "wheels" of this diagram together. One, for example, is an active, pupil-centred approach to learning: a second the issue of making science relevant to the pupil and to society. The reader will no doubt find many others.

So the book attempts to bring together within one set of covers, views, information, advice and help on a spectrum of issues relevant, in the extended professional sense, to the secondary science teacher today: and to structure this material in such a way that readers can open the book at a place appropriate to their current concerns. The authors are all experienced science educators and teachers, and bring their own knowledge, experience, expertise and enthusiasm to their chapters. Where appropriate, they have provided guidance, in the form of checklists or activities, to assist the reader in thinking about the issues, to help in policy making and implementation, and to address practical matters.

It is hoped that this book will help ever more busy science teachers to tackle an ever more complex job.

Richard Hull

August 1992

Richard Hull was formerly a Lecturer in Science Education in the School of Education, Nottingham University, and is now freelance. He is a member of ASE Validation Board and has written books on teaching science and the organisation of science departments.

Setting the scene

1 The Purposes of Science Education

Paul Black

1 Why Discuss Purposes? – an introduction

It is not inevitable that science be given the high priority that it now enjoys in the school curriculum, or even that it be a subject at all. A case could be made out, for example, that medicine, or politics should be essential components. Thus one priority in a discussion of purposes is to justify the place of science in schools. Why, for example, should science be required in both primary and secondary stages, rather than, as for languages, at secondary only? And does it have to be a separate subject, when other aspects of education are deemed worthy only of cross-disciplinary or extracurricular attention, notably personal and social education and environmental education?

The orientation and practice of science education have to be consistent with the justifications given for its place in the curriculum. So a second function of any discussion of purposes is to set out the aims which science education should deliver and in the light of which it should be evaluated.

There could be a cynical response to the inevitably general and idealistic tone of this chapter. We are living in National Curriculum country (DES 1991) and whilst a trip to an imaginary island could provide entertainment and relaxation, it might also generate frustration on return. One justification here is that there is a wide range of classroom interpretations of the curriculum and it is in the classroom that each teacher's personal hopes strongly influence their children's experience.

A second justification is that the motivations behind a formal curriculum document need to be understood by anyone who has to put it into practice. Many of the issues discussed in this chapter receive no explicit mention in the national curriculum order, yet teachers' views about these issues must influence their practice. This point leads onto a more general justification – curriculum thinking is a product of its time and is open to change. We all need to be prepared to contribute to change in the future, and not just to be victims of its tides (ASE 1992)

This last issue is taken up in the next section of this chapter, which surveys the historical context, and in section 3, in which the factors that might influence change are reviewed. Following these preparatory discussions, section 4 discusses the main aims of science education for all, whilst Section 5 discusses the needs of future specialists. These lead to an outline, in section 6, of a proposal for the main elements of science learning which could implement the purposes. Section 7 then takes up the problem of the place of science in the whole school curriculum. Section 8 concludes with discussion of some implications.

Little has been said here of the concept of "scientific literacy" (Champagne and Lovitts 1989, Atkin and Helms 1992). It could be said that the phrase stands for no more than the science needed by all. However, the use of the metaphor of "literacy" might be a way of giving a particular emphasis, giving priority to narrowly instrumental needs. In short, the phrase either adds nothing to the arguments here, or is used to introduce, by implication, a bias of priorities. Therefore, it does not seem to be helpful. (See p 361 for other discussion of the idea).

2 How Did We Get Here? – the historical context

The nineteenth century beginnings of science education in schools have been described in David Layton's book *"Science for the People"* (Layton 1973). Conflict arose between two groups holding opposed views about the purposes of the new development. One proposed a science that would focus on everyday practices and artefacts. The second wanted to focus on academic science so that school science would help to recruit future scientists. The second view prevailed then and has done so ever since, although that original tension continues to bedevil school science to this day.

Whilst such influence dominated in the first part of this century, there also emerged educational reformers, notably Armstrong, who had a vision of a change which would give pupils a more authentic engagement with the practice of scientific enquiry (Jenkins 1979). However, the style of science teaching remained largely formal, based on teaching of definitions and derivations, and on experiments which illustrated foregone conclusions, with some discussions of applications added at the end.

This tradition was transformed in the 1950's and 1960's by a worldwide movement for reform, driven by and originating from two main foci. One was in the United States, where fears of technological inferiority sparked off by a Russian lead in space flight served to generate support for curriculum reform. The other was in Britain; here, it was the dissatisfaction of science teachers themselves which led them to press for reform. However, the outcome, in the Nuffield Foundation's science teaching project, was supported and influenced strongly by university scientists.

The assumption in the USA movements was that better school curricula would lead to better prepared students for science and engineering. In consequence, the courses were "top-down" in their construction, showing originality in devising quite new routes to the understanding of fundamental concepts. The agenda of these was chosen to represent current science in an authentic way, an aspect in which existing courses notably failed. This new strength was also a weakness, for the courses were ambitiously abstract and paid little attention to everyday consequences. A later physics project, *Project Physics*, differed from the others in giving more emphasis to the history of the development of ideas and to showing how they developed within the societies – and technologies – of their times. Subsequent evaluations showed that on the whole these courses, and the many international applications and imitations which they fostered, did not achieve their aims (Tamir et al 1979).

The Nuffield courses were more successful in holding those originally committed. The first courses were designed for the most able and were thereby biased to academic science, again giving priority to clearer emphasis on modern concepts rather than to any broader perspectives on science. They were characterised however by more emphasis on changes in teaching method. A second generation of Nuffield courses struggled with a different problem, the provision of courses firstly for those placed in streams or schools designed for the less able, later for the mixed ability groups of the secondary comprehensive school.

The implementation of comprehensive schooling led to new challenges, as it became increasingly clear that quite new arrangements for science teaching would be needed. Growing dissatisfaction, with courses that were too specialised, and with the need for pupils to opt out of one or more of the separate science subjects, culminated in the 1980's in a move for broad and balanced science, supported by the ASE, the DES and, notably, by the Secondary Science Curriculum Review. (ASE 1981, SSCR 1987)

These changes were motivated by a desire to realise the purpose of making science education accessible and meaningful to all pupils. However, this needed more than a change in the arrangement of courses. Three further aspects were given a powerful impetus in the 1980's. One emerged from research studies reporting both that pupils' enthusiasm for science declined from a high point of enthusiasm at age 11 to a low level by the end of schooling, and that only a small minority attained any satisfactory grasp of even the most elementary concepts by age 16 (Driver et al 1985, Osborne and Freyberg 1985).

The second trend started from the new emphasis given to the processes and skills of science, particularly through the work of the APU science monitoring (Black 1990). This led to more complex views of the essential links between the learning of concepts and of skills. These developments also led to a new emphasis on pupils' own science investigations, culminating in the emergence of Attainment Target One of the national curriculum.

The third trend started with the growing concern with environmental issues and with the moral responsibilities of scientists, which led some to promote the study in school science of its technological and social implications (Lewis 1980, Solomon, 1983, Holman 1986, Hunt 1988). (See Chapter 20). Others argued for a more human approach which would discuss scientists as persons developing their contributions in particular historical contexts. A third group criticised science teaching for failing to give explicit attention to the methods of science, and yet presenting, by implication, quite false views about them (Hodson 1985, Lucas 1990). These strands have converged in arguments that a broader, more humanistic, definition of school science is now needed (Matthews 1990, 1992).

The position is no less stable in the area of primary science (Black 1985). The early predominance of the "nature table" approach was first replaced by attempts to set up a broader agenda of interesting topics for basic knowledge. This was further replaced by an emphasis on process skills, with reduction or even exclusion of specific topic knowledge. This has been replaced again, notably in the national curriculum, by a model which emphasises investigative skills, together with a mixture of basic topics ("know that" statements) and elementary concepts ("understand that").

The sum of these trends has been a definition of the science curriculum for the present decade which is different in several fundamental respects even from that of the early 1980's. Indeed, there have been fundamental changes in every one of the last few decades.

3 Stop the World We Want to Get Off! – the future for change

It might be comforting to imagine that future decades will be marked by greater stability. However, it seems unlikely that the forces which have driven the changes of past decades will disappear. The first of these is the influence of changes in science and technology themselves. Areas such as genetic engineering, the AIDS epidemic and the burgeoning environmental crises, let alone others that are yet to emerge, must surely affect priorities in science education.

The second drive for change will come from changes within education itself. In the last ten years, research in education has become notably more relevant to pedagogy than at any previous time, and this process is hardly likely to stop. At the same time, developments in information technology have yet to achieve anything like the impact which is currently foreseeable. For example, when vast resources, in knowledge bases and training systems, are available to anyone at the touch of a button, what then will be the point of much of the work that now takes up time in classrooms?

The third drive will come from changes in society, leading to changes in expectations for schools. The national curriculum itself is not an accident arising from the 1987 general election, but rather the culmination of longer term political and social trends. There is no sign that public and political interest in education has since declined, and no evidence at all that the definition of a national curriculum will be followed by suspension of debate about curriculum purposes or practices.

Finally, there is the less evident but most potent catalyst, changes in pupils themselves. Radical changes in influences outside the school, notably in family life and in the media may well be producing profound changes in the needs of the pupils. A recent authoritative review of the condition of children in the USA (Hamburg 1992) talks of a generation in crisis, because of the disastrous stresses for those growing up in poverty in inner cities, and the neglect of parenting in all classes. As society starts to suffer the consequences of such changes, it may look to schools to provide a redeeming framework for many children – a task which far transcends their traditional role.

Given all this, it is up to teachers and schools to develop to the full their understanding of, and so their commitment to, their purposes for education. They might thereby be better able to hold on to their own interpretations of change, so that they can filter or distil the pressures in the interests of their pupils, rather than be disoriented and de-skilled by them. They might also be better prepared to influence public debate about re-definitions of their purposes and priorities.

However, the influence of teachers is bound to be limited. Others professionally involved in science education, whether as academic researchers, as inspectors, as trainers or as administrators, will all exert pressure. The growing emphasis on the need to expand the numbers in tertiary education will give that sector more influence. More uncertain will be the strength of public, notably political pressure. Business, industry, and environmentalists may play enhanced roles, whilst it seems that political control over the curriculum, far from leading to stability, is leading to more questioning of school practices and to weakness in the face of the temptations to exert one's powers to "improve" as soon as difficulties are publicised.

Thus, the determination of purposes and priorities for the science curriculum has to take place in an essentially political struggle between competing traditions, perspectives and interests. In this respect, it does not differ from other significant social issues. However, the point deserves emphasis here, both because current trends and proposals have to be examined in this light, and because formulation of policies and practices in the future have to be made in this context. However, whilst schools are not in full control of specification, they have very strong control over implementation, so that their own priorities, and their views of the priorities of others, are of outstanding importance.

4 Science for All?

The starting point must be to consider the purposes of a curriculum which might be the only experience of serious learning about science that pupils have in their lives. Given the large and growing relevance of science in the private, social and political spheres, the optimum planning of this experience must be of the utmost importance.

None of the main purposes can be achieved unless certain subsidiary purposes are met. The first of these is *accessibility*. Pupils must understand and feel confidence with the science they are studying. This purpose may present very difficult dilemmas, for it is clear that in experience to date, it has not been achieved .

Likewise, the main purposes cannot be achieved unless pupils can see the *relevance* of what they are trying to learn and can find stimulus and *enjoyment* in it. Very few pupils can persevere with work which, being difficult and apparently irrelevant, brings little immediate reward.

To turn to the main purposes, the first is that pupils should be given *a basis for understanding and for coping with their lives*. Science has a lot to say about problems in people's personal lives, notably in health, including nutrition, drug abuse, the *AIDS* epidemic and in broader issues, such as those concerned with sexuality. Here, as for most applications of science in personal and social life, impersonal knowledge and understanding has to develop in conjunction with appreciation of issues of moral and social value. This purpose embraces the need to look after oneself, and to help protect oneself and others from the flow of incomplete and misleading information which seems to be an inevitable feature of a democratic society.

The need to become capable of taking part in positive initiatives to improve life for all follows naturally. Indeed, the social dimension expands this agenda to such an extent that a further purpose, that of *understanding the applications and effects of science in society* is worth formulating. The considerations here also need to be related to issues of value and morality, and also raise the question of whether such issues should be tackled only, or even mainly, in the science classroom. Almost any of the main examples, for instance pollution or global warming, could well be tackled from a variety of non-science perspectives. This point is taken up further below.

The two aspects of the purpose discussed so far involve learning *from* science, using its results as a starting point, rather than *about* science, studying how those results are achieved. Learning about science constitutes the second main purpose. It involves learning about *the concepts and the methods which are combined in scientific enquiry.* Pursuit of this purpose in isolation can give too much emphasis to purely instrumental reasons for learning science. One function of schools however is to be the guardians and transmitters of a society's culture. Science is one of mankind's greatest achievements, and without some knowledge of its history, some appreciation of the personal genius of famous scientists in determining the course of science, and some insight into the particular way in which science searches for truth, a pupil will have little *insight into what science, seen as a human activity, is really like.* For example, a pupil struggling with the ideas that air fills the space around her, and that it has weight and that it even exerts a very large pressure, might understand her own confusions and the nature of science better if she was told about how some of the greatest intellects grappled with the same difficulties in previous centuries. Historical examples can also make clear that imagination and creativity are essential to the development of science, and that it proceeds through obstacles of misunderstanding, and controversy, rather than by smooth deductions and experimental "proofs".

Such purposes should be so pursued, that the experience of science that school learning gives will be *authentic*. This need affects the style of the work undertaken, the range of issues covered and the image of science that is conveyed.

Science education cannot be planned as an isolated experience, contributing only to its own particular purposes. Thus a third main purpose is that *it must contribute to the general personal and intellectual development of the pupils*. Such contribution tends to be considered automatically in primary education, but to fall away at secondary level where the practice of subjects working in isolation is too prevalent.

There are many possibilities here. One is that a pupil should gain experience in science of the *logic of explanation* using assumptions, models, evidence and argument to reach conclusions. Such experience can build up pupils' confidence in their power to construct explanations. A different emphasis appears in the possibility of pursuing practical investigations. These can contribute to building up practical capability, since such investigations call for powers of initiative, of making decisions, of overcoming obstacles. They can also provide outstanding opportunities for learning to work in small groups. For both of these aspects, it is essential that emphasis be given to helping pupils to reflect on what they have done so that they become more aware of the way in which science works (White and Gunstone 1989). Here, the pursuit of the goal of authentic understanding of the nature of scientific enquiry contributes powerfully to pupils' general development. Such contribution can be enhanced if pupils can relate the way in which they work in science to the ways in which they work in other disciplines.

This agenda cannot be complete without attention to the fact that a common curriculum should provide pupils with *a basis for making choices, together with positive motivation* to consider seriously a further commitment to science. This, the fourth main purpose, is discussed in the next section.

5 Your Country Needs You! – providing future specialists

Given the spectrum of purposes set out above, it can be asked whether a curriculum which gave its main priority to the selection and preparation of future scientists ought to be very different. The needs of understanding science, with its implications for developing powers of explanation and of practical capability, are of direct significance for future specialists. It can hardly be argued that a broader view of science as a human activity is not needed by those in whose hands the future development of science will lie. Indeed, it can be argued that the neglect of this aspect at all levels of study, particularly in specialised tertiary education, has been an intellectual weakness and a practical impoverishment of the scientific community. The need for the experience of science to be authentic would seem to be a top priority in providing pupils with a basis for choice. It might be argued that a far narrower range of subjects, studied in greater depth, and giving less attention to the broad range of everyday applications, might be appropriate. This raises the prospect that future specialists would be less well served than their peers in using science to cope with their personal and social needs.

However, it is also necessary to consider the need to attract pupils to specialise in science as well as the need to prepare them if they are attracted. The number attracted has always been too small and, for science and engineering, has always included far too few girls. This means that the images and experiences of science presented to school pupils have to be changed. Studies of the links between adolescent personality development and choice of science (Head 1986) indicate that the closed and algorithmic view of science that courses usually present tends to repel those extrovert pupils, particularly boys, who wish to challenge authority and question values. For girls, the impersonal view of science presented by a narrow concentration on "basics" is a severe obstacle. Thus, emphasis on a narrow preparation can limit recruitment. It might have the further disadvantage that those who are attracted are being misled, because the reality of professional practice will be quite different from their experience of "preparation".

The conclusion from these arguments is that future specialists cannot do without any of the work that all pupils require. However their needs argue for the provision of more in-depth work in at least a sample of areas of science, so that any pupils who wish can explore their commitment and test out their ability to take science further. This particular part of the argument has no relevance for primary science. However, insofar as primary science generates enthusiasm, an appreciation of science as the work of people, and a growth in capability and in understanding it is laying essential foundations for all of the purposes discussed as well as providing the first inspiration for future scientists.

Evidence that students in tertiary education suffer from the same misconceptions about science fundamentals as their secondary peers, and that transfer of learning to new contexts is notoriously difficult to achieve, cast doubt on the notion that school study can at present provide a "firm grounding". A counter-argument, that a spiral curriculum in which a broader range of aims is addressed at each stage may be more effective, is at least plausible. Thus, for example, an interplay between principles, applications, and "hands-on" investigation should characterise both sides of the secondary – tertiary interface, just as it should at the primary – secondary interface, for which it would be absurd to propose that primary pupils should learn only concepts, leaving project investigations to build on this "preparation" at secondary level.

The arguments may be different if the aim is to provide pupils with preparation that is more directly vocational. Arguments about the needs of different vocational groups are often in mutual conflict, so that it emerges that the range of needs can only be met by distilling out what they have in common. Thus the argument reduces to one about basic knowledge and skills. The metaphor of "basic skills" has many attractions. There are dangers here also. The implied assumption, that these skills are the same across different contexts, needs careful examination; this assumption may be evident in the case of reading instrument scales but it is very doubtful if (say) problem-solving is assumed to be the same skill in many different occupations. It is also notorious that pupils are usually unwilling or unable to apply a skill, learnt in one context, to a problem in a quite different context.

Finally, it should be noted that some parts of higher education are also calling into question the view that specialist education should proceed by the learning of required components at early stages and should only tackle the "synthesis-in-application" of these at advanced stages. There is, for example, a move in engineering schools in the USA to emphasis the integrative and holistic activity of tackling real problems right from the start of freshman courses (Bordogna 1989), whilst science degree courses in this country now incorporate substantial components of project work where thirty years ago they would have assumed that such work had to wait until the postgraduate stage.

6 A Model for Purposes – essential components of a learning programme

A brief discussion is offered here about components essential to any learning programme designed to achieve these purposes, partly because this will help to bring out some of the implications of the aims presented.

A first essential is that students should come to understand science and to understand how science is made, by being engaged in doing it. This involves three main aspects. One is that they have to learn about its main concepts, seen as abstract yet powerful agents for predicting and controlling natural phenomena. This in itself is a formidable requirement. The difficulties pupils of all ages have in grasping and accepting many science concepts have been well documented. It seems clear that to overcome this problem requires that more classroom time be spent on any one idea, so the range of concepts to be covered will have to be reduced (Scott et al 1992).

The second aspect is that pupils should be able to use the main skills which go to make up the scientific method. Such skills as observation, measurement, making generalisations, inventing hypotheses, devising fair tests, designing experiments, analyzing data and interpreting results should all be included. A full exploration of many of these has been made in the work of the science teams of the APU (Black 1990, Strang 1990, Strang et al 1991)

However, neither the concepts nor the skills can be properly understood in isolation from one another. It is not possible, for example, to propose hypotheses except in the light of some preconceived model of the system under consideration. Conversely, understanding of the ways in which concepts have developed can only be conveyed through activity which uses these skills. This leads to the third aspect – which is that pupils should have personal experience of working with the interaction of concepts and skills in planning, designing, carrying out and interpreting their own experiments. Only through such activity can pupils develop an authentic understanding of what is involved in doing science. It is important that at least some of these should involve the application, and subsequent modification, of scientific ideas to test explanations. Science involves far more than systematic comparison of materials or products, and whilst consumer tests may have some value in teaching some skills, some published materials so emphasise this approach that they give a misleading image of science.

It is not implied that every new idea must be developed through students' own personal investigations. However, their own personal experience of investigating some ideas and phenomena should help pupils in understanding the way in which others have come to new ideas through scientific investigation.

Whilst it should be clear that the careful pursuit of these three aspects in an interconnected way must be the main vehicle for achieving some of the main purposes, other activities will be needed to complete the agenda. Students should come to know about science as a human activity, by studying how its achievements have emerged in particular historical contexts and have depended on beliefs, technology, social systems and above all the human personalities of those involved. Reflection on their own investigations should help pupils to develop understanding of the ways in which evidence, experimentation, hypotheses, models and mathematics are combined in the development of science.

The above requirements demand that more time be spent on some science topics, because there is evidence that unless this is done we shall continue to leave pupils with serious misunderstandings, not only about particular scientific ideas, but also about what the whole activity of science is about. However, this need conflicts with the purposes of giving pupils a basis to cope with everyday needs, for this requires that they know about the many results of scientific work which are important to them in their lives. It seems neither possible nor necessary that all of this large number could be studied in the depth suggested in the tripartite interaction of concepts, skills and processes discussed above. Whilst a few must be studied in this way, others could be studied more superficially, with emphasis on those features which must be known about in order to lead a healthy and responsible adult life.

The need to know about the many applications of science, and about the ways in which, through its contribution to technology, science has had a profound effect on our society, is indisputable. However, it is difficult to specify what level of understanding of a scientific idea is needed to be able to make use of it in daily life (Layton 1991): electricians know far less about circuit theory than physicists, but they are better at fixing a new ring main. At a more general level, the extent to which science has been the driving force of modern technology is often exaggerated, there being many technological changes in which the contrivance technology of craftsmen and entrepreneurs has been the main driving force (Gardner 1992). Finally, since technological change is a complex human activity in which many areas of thought and action play a part, it may be misleading to present it to pupils in the restricted context of the science classroom (Black and Harrison 1985). This point will be taken further in the next section.

All of this discussion has treated science as a single entity. The conceptual basis of science has a structure in which the separate components of biology, chemistry, physics and earth sciences are identifiable, albeit overlapping, components. Furthermore, these different areas of science differ significantly both in philosophy and in their styles of work so, for example, biologists overlap with social scientists in the way that physicists do not, and biologists have to do experiments with many of the variables uncontrolled, the design of which would be unthinkable for a physicist (Black 1986).

Opinions differ about the implications of these features. Several varieties of combined, co-ordinated, or integrated science are used. Almost all agree that to teach three or four sciences separately and without close co-ordination is unacceptable and that up to age 16 all the main areas should be encountered by all. It is also the case that, in almost all courses, whatever their ideology, there is a mixture of separate discipline topics and topics which clearly cross the boundaries. These features contribute to an overall purpose of understanding the nature of science, which has within itself elements both of unity and of diversity.

7 Across the Curriculum?

If a broad definition of technology is adopted, then it has a place of its own in the curriculum which does not derive from science. In this view, one problem about implementing technology in the curriculum is to ensure that, in their learning and involvement in technology, pupils summon up and bring into use contributions from many school subjects, of which science is one. Science teachers should try to make their contribution to technology, both to serve other parts of the curriculum, and give their own pupils a mature appreciation of the role of science in technology.

There is a broader set of possibilities to which similar considerations apply. Science is an arena in which the purposes of other important parts of the curriculum can be realised, e.g. uses of descriptive and imaginative language, or application of mathematical modelling, or development of moral education in relation to the choices presented by science. Conversely, it can be enriched by work in other subjects. Examples are overlaps between earth sciences and geography, and overlaps with history; the latter would be important if the scope of science study were to be broadened to make it more human (Watts 1991).

Such opportunities are taken up in primary schools, although it needs careful planning if topic work is to bring out the inter-connections. There are more serious difficulties at secondary level. Where school subject departments work as separate empires, the pupils are left to make the inter-relationships which their teachers have failed to make for them; most cannot do so and keep their learning of subjects in separate compartments. This grave weakness can be overcome if schools give priority to planning the curriculum as a whole. This requires that overarching themes be formulated and agreed, from which the roles of separate subjects can be assigned and the possibilities of inter-subject work explored.

Sadly, the national curriculum, with its almost exclusive concentration on separate subjects and avoidance of general curriculum principles or aims, gives too little encouragement to such important work.

8 Making it Happen

The realisation of purposes in classroom work is notoriously difficult. Of course, where discussion of them is treated as academic indulgence without any vision of accepting the struggles needed to make any important changes, there can be little point in discussing purposes at all. If the possibility that they are to make a serious difference is accepted, then one approach would be to carry out an audit of present activities to judge which are being served and where gaps occur. This would then lead to difficult decisions as to whether the purposes can be more fully met by a set of small adjustments, or whether quite radical changes are needed, at least in part.

A helpful part of any such planning is to derive from the broad aims some secondary aims which necessarily follow from them; for such aims, being closer to practice, can help cross the bridge between high-minded purposes and day-to-day work. Some of the discussion in Sections 6 and 7 can be read as contributions to such a strategy. Within such a strategy, the issues of progression, implying the matching of the purposes to the development of pupils with age, would also need consideration. This has not been attempted here, apart from a few references to the differences between primary and secondary stages. (But see Chapter 15). It is not proposed that any of these purposes should be uniquely reserved for one phase only – they all apply at most ages, albeit in different ways.

Realisation of new purposes can imply more than shuffling of lesson plans so that new topics are introduced or the old ones realigned. For example, if moral issues are to be taken up, discussions which cannot lead to the right answer have to be managed; if pupils are to learn from taking responsibility for their own experiments, the role of their teacher will be quite different from the role needed to guide routine practical exercises (Black et al 1992). Such role changes are the most difficult obstacles in the path of any large shift in the purposes of a curriculum.

Finally, success might also require that pupils be aware of the underlying plan and perhaps build up their own portfolio of ideas about science and about the point of studying it (Claxton 1990). Central to any debate about purposes is the need to achieve reconciliation between what is needed by all, what is wanted by all and what is feasible for all. What is wanted by all is an oft neglected aspect.

What is feasible for all is the great uncertainty. Wherever new purposes, or a radically different balance between purposes, are explored, implementation changes the view, not only of what is feasible, but of what is really desirable (Roberts 1988). Commitment to a new vision of purposes is a moral adventure – a voyage in which the purposes themselves are explored as well as the uncharted tracks that lead to them.

Paul Black is Professor of Science Education at King's College London

References

References are given in the sections above where they are relevant to particular points and a list of these is given below. However, for more general discussions of purpose, a few of those listed are particularly valuable – these are identified by asterisks.

*ASE (1981) *Education through science:* Policy statement, ASE, Hatfield.

*ASE (1992) *Change in our future; a challenge for science education,* ASE, Hatfield.

Atkin J M and Helms J (1992) Private communication, Stanford University, California.

Black P and Harrison G (1985) *In place of confusion,* Nuffield/Chelsea Curriculum Trust, London.

Black P J (1990) *APU Science: the past and the future,* School Science Review, Vol 72, no 258, pp 13 – 28.

Black P J, Fairbrother R, Jones A, Simon S and Watson R (1992) *Open work in science: a review of practice,* King' Research Paper, CES, King's College London. (See also Black P J, Fairbrother R, Jones A, Simon S and Watson R, *Development of open work in school science*, a book to be published by the ASE.)

Black P J (1986) *Integrated or co-ordinated science?* (Presidential address given to Association for Science Education at University of York on 4th January 1986), School Science Review, Vol 67, no 241, pp 669-681.

Black P J (1985) *Why hasn't it worked?* Chapter in: *Approaching primary science* (ed B Hodgson & E Scanlon), Harper & Row, London, pp 61-64.

Bordogna J (1989) *Entering the 90's: A national vision for engineering education,* Engineers' Education vol 79, no 7.

Champagne A and Lovitts B (1989) *Scientific literacy: a concept in search of a definition.* In Champagne A, Lovitts B, and Calinger B (Ed's), *This year in school science 1989: Scientific literacy.* Papers from the 1989 AAAS forum for school science, AAAS Washington DC, USA

Claxton G (1990) *Science lessens?,* Studies in Science Education 18, pp 165-171.

DES (1991) *Science in the National Curriculum 1991,* HMSO.

Driver R, Guesne E and Tiberghien A (Ed's) (1985) *Children's ideas in science,* Open University Press, Milton Keynes U.K

*Fensham P J (Ed) (1988) *Development and dilemmas in science education,* Falmer, Lewes U.K.

*Fensham P J (1992) *Science and technology,* pp 789-829, in *Handbook of Research on Curriculum,* Macmillan, New York.

Gardner P L (To be published) *The application of science to technology,* to be published in *Research in Science Education* – papers from the 1992 ASERA conference at Waikato University New Zealand.

Hamburg D A (1992) *Today's children: Creating a future for a generation at risk,* Times books – Random House, New York.

Head J (1986) *The personal response to science,* Cambridge University Press.

Hodson D (1985) *Philosophy of science, science and science education,* Studies in Science Education 12, pp 25-57.

Holman J (1986) *Science and Technology in Society, General guide for teachers,* ASE, Hatfield.

Hunt A (1988) *SATIS Approach to STS,* International Journal of Science Education, 10, pp 409-420.

Jenkins E (1979) *From Armstrong to Nuffield,* Murray, London.

Layton D (1973) *Science for the People,* Allen and Unwin, London.

Layton D (1991) *Science education and praxis: the relationship of school science to practical action,* Studies in Science Education 19, pp 43–78.

Lewis J (1980) *Science in Society: Readers and teachers guide,* Heinemann, London.

Lucas A M (1990) *Processes of science and processes of learning,* Studies in Science Education, 18, pp 172-177.

Matthews M R (1990) *History, philosophy and science teaching: a rapprochement,* Studies in Science Education, 18, pp 25-51.

Matthews M R (1992) *History, philosophy and science teaching: The present rapprochement,* Science and Education, 1, pp 11–47.

Osborne R and Freyberg P (1985) *Learning in science: the implications of children's science,* Heinemann, Auckland N.Z.

*Rutherford F J and Ahlgren A (1989) *Science for all Americans,* Oxford University Press, Oxford.

Roberts D A (1988) *What counts as science education?* pp 27–54 in Fensham P J (ed) *Development and dilemmas in science education,* Falmer, Lewes U.K.

Scott P H, Asoko H M and Driver R (1992) *Teaching for conceptual change: a review of strategies,* in Duit R et al (Ed's) *Research in physics learning: theoretical issues and empirical studies,* Kiel: IPN – Institute for Science Education.

*SSCR (1987) *Better science: making it happen,* ASE/Heinemann, London.

Solomon J (1983) *SISCON in schools – Readers and teachers guide,* Blackwell, Oxford.

Strang J (1990) *Measurement in school science.* Assessment matters No 2, Schools Examinations and Assessment Council, London.

Strang J , Daniels S , and Bell J (1991) *Planning and carrying out investigations.* Assessment Matters No 6, Schools Examinations and Assessment Council, London.

Tamir P et al (1979) *Curriculum implementation and its relationship to curriculum development in science,* Israel Science Teaching Centre, Jerusalem.

Watts M (Ed) (1991) *Science in the National Curriculum,* Cassell, London.

White R and Gunstone R (1989) *Meta-learning and conceptual change,* International Journal of Science Education, Vol 11, no 5, 577-586, 1989.

2 A National Curriculum for Science

David Oakley

1 Why a National Curriculum?

1.1 The background

1.1.1 The political imperative

The National Curriculum (NC) was introduced as a result of the reforming philosophy of a Conservative government which culminated in the Education Reform Act of 1988. ERA contained a mass of legislation (nearly 1 cm thick) which gave the Secretaries of State for England and Wales (SoS) unprecedented powers over the education system in this country. As well as the legislation for the National Curriculum, it included that for Local Management of Schools (LMS), open enrolment, the establishment of the National Curriculum Council (NCC), the Curriculum Council for Wales (CCW) and the Schools Examinations and Assessment Council (SEAC). Many objected at the time to the setting up of separate organisations for curriculum and assessment, and indeed the government white paper of August 1992 proposes to combine them again in the School Curriculum and Assessment Authority (SCAA).

In 1989 the Secretary of State (Kenneth Baker) summed up the aims of the National Curriculum (in a speech at the North of England Conference) as:

1 giving a clear incentive for all schools to catch up with the best and the best will be challenged to do even better;
2 providing parents with clear and accurate information;
3 ensuring continuity and progression from one year to another, from one school to another;
4 helping teachers concentrate on the task of getting the best possible results from each individual child.

All state schools in England and Wales are required by law to provide a balanced and broadly based curriculum which will prepare young people for the opportunities, responsibilities and experiences of adult life and promote the spiritual, moral, cultural, mental and physical development, of pupils at the school and of society. Fears about the prescriptive nature of a National Curriculum were deflected by statements that it is a "framework not a straightjacket" and is a "minimum entitlement". How the "Programmes of Study" (see p 31) were to be taught and learned was not specified, though government concern about teaching methods resulted, in 1992, in the setting up of a three-man commission (Alexander, Rose and Woodhead) to consider primary practice, and in the setting of new guidelines for a reduced role for course-work at GCSE. Both steps encouraged debate about how pupils should best be taught, and challenged aspects of current practice.

1.1.2 A view of the curriculum

The curriculum model presented was subject-dominated, with ten subjects plus Religious Education to be accommodated. Crosscurricular themes, dimensions and skills were added later, part-way through the process of introduction of the subjects. Science was deemed to be of fundamental importance and was elevated to the "core" curriculum alongside English and Maths. The recognition of science as an essential part of the curriculum represented a major success for those influencing national policy on education. The seminal HMI document *"Science 5–16:A statement of policy"* (DES 1985) the ASE and the Secondary Science Curriculum Review, all played a part in preparing the ground for the core curriculum status for science, as did the Education Support (ESG) Grants to help LEAs' to strengthen the primary science curriculum.

1.1.3 The making of the curriculum

Once the government had embarked on this course of action, someone had to advise on, and write, the curriculum. The first group set up to pave the way was the Task Group on Assessment and Testing (TGAT). Chaired by Professor Paul Black of Kings College, London University, this was influential, in particular, in suggesting the ideas of attainment targets(AT), the ten levels of attainment (see p 30 for definitions) profile components, (groups of attainment targets for reporting purposes) and standard assessment tasks, for national testing of achievement. Many of their original ideas have evolved; some have fallen by the wayside. With TGAT to give guidance, working groups were set up by the Secretaries of State to produce recommendations on what the curriculum should be. The science working group was chaired by Professor Jeff Thompson of Bath University: academics and practitioners were represented, and HMI and NCC officials attended the meetings. The remit of the group also included Technology 5–11. The timescale was short. They met at regular intervals over a twelve month period. Their early thoughts were summarised in an interim report in December 1987. The thinking of the group produced, from five "themes", a number of attainment targets, 22 in all in their final report, reflecting not only subject content areas but also skills. The attainment targets (ATs) were packaged into profile components as a way of grouping them for reporting purposes. Consultation conducted by the NCC resulted in recommendations to the Secretaries of State and in March 1989 Draft Orders were framed, followed, after brief consultation, by Final Orders, which itemised 17 Attainment Targets grouped into two profile components. Only 14 ATs were thought to be appropriate for pupils from 5–11. 4 ATs were proposed for Primary Technology. Fig's 1 and 2 summarise the sequence of events and the "evolution" of National Curriculum Science. One notable feature in fig 1 is the changing balance between knowledge and understanding and skills.

1.1.4 Implementing the National Curriculum

Schools were required to start to implement NC Science, from August 1989 for five year olds and twelve year olds, and from August 1990 for eight year olds. A massive in-service training exercise was mounted by LEA's to prepare teachers for delivery of the new curriculum. Fortunately, many primary advisory teachers were in post, funded by Education Support Grants and the Secondary Science Curriculum Review had stimulated the appointment of secondary advisory teachers. Advisory teachers were recruited from amongst good classroom practitioners to lead in-service training, to prepare support materials for teachers and co-ordinators, to work alongside teachers in the classroom and act as consultants and advisers. Many were seconded to the role and returned to the classroom; others became education lecturers or advisers/inspectors. With the demise of ESG's, the rise of LMS and the devolution of finance to schools, the number of advisory teachers has declined, but their support was crucial at the time.

1.2 The issues

In primary schools, the immediate task of teachers was to come to terms with National Curriculum in Maths, English and Science and incorporate them into the curriculum. The emphasis on subjects initially caused problems, as a result of the apparent conflict with the integrated nature of the topic work characteristic of primary practice. Guidance focused on detailed planning for coverage of the Programmes of Study (PoS) and attainment targets over a year and a phase. Overemphasis on Attainment Targets tended to spoil the coherence of some of the topic work by narrowing its focus, and it took a while for teachers to realise that ATs are not "taught" but attained. This also proved to be true for secondary schools. Even though the NC was to be

	Attainment Targets	Profile Components	Weighting of PCs %				Examples

	Attainment Targets	Profile Components	KS1	2	3	4	Examples
Interim report December 1987	5 "themes" define the scope of science • living things and their interaction with the environment • materials and their characteristics • energy and matter • forces and their effects • the earth and space						
Final report & proposals of SoS August 1988	1 – 16	knowledge and understanding	35	35	40	40	No
	17 & 18	exploration and investigation	50	50	30	25	
	19 & 20	communication	15	15	15	15	
	21 & 22	science in action	–	–	15	20	
NCC consultation & recommendations Dec 1988 and draft orders January 1989	1	exploration of science	50	45	35	30	No
	2 – 17	knowledge and understanding of science	50	55	65	70	
Statutory orders March 1989	1	exploration of science	50	45	35	30	No
	2 – 17	knowledge and understanding of science	50	55	65	70	
Proposals of SoS May 1991	1	exploration of science	50	45	35	30	
	2 – 5	knowledge and understanding of science	50	55	65	70	
NCC consultation & recommendations September 1991			weighting of ATs				
	1	no profile components	50	50	25	25	Yes
	2 – 4		50	50	75	75	
Draft orders October 1991			weighting of ATs				
	1	no profile components	50	50	25	25	No
	2		} 50	} 50	25	25	
	3				25	25	
	4				25	25	
Statutory orders 1991	1	scientific investigation	50	50	25	25	Yes
	2	life & living processes					
	3	materials & their properties	50	50	75	75	
	4	physical processes					

Figure 1 The evolution of the National Curriculum for Science.

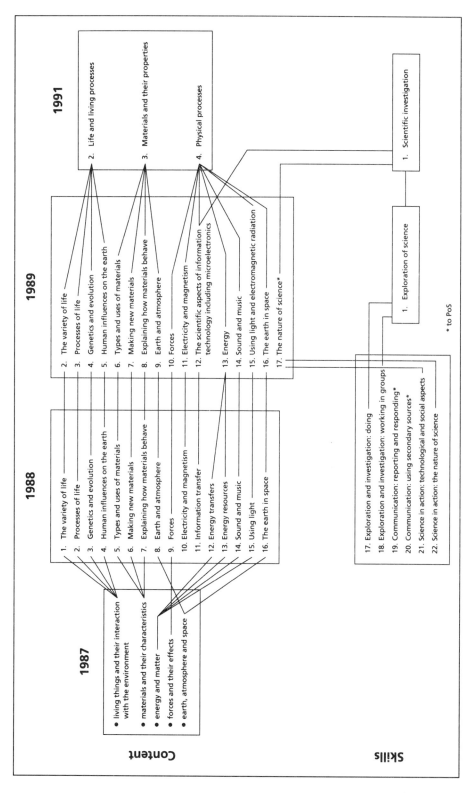

Figure 2 National Curriculum Science: Content and Skills Attainment Targets 1987-1991

phased in gradually, primary schools soon realised the value of implementation in all year groups, to allow more coherent planning, particularly as the NC provided more of a framework for continuity and progression, and offered more specific guidance on what pupils at a particular age should be capable of, than existed before. It was no longer sufficient for individual teachers to plan in isolation from their colleagues and disregard what had gone before and what came after. This has perhaps been one of the greatest benefits of the National Curriculum.

1.2.1 Key Stage 1
At KS1, (see p 32 for the meaning of "key stage")) teachers were able to provide a much better and more consistent and complete coverage of science. For many, the problems of lack of funding and the speed of curriculum development were more than balanced out by the sharpening of their own practice, although detailed planning to some extent reduced the spontaneity, flexibility and serendipity of the early years.

1.2.2 Key Stage 2
KS2 presented the additional problem of the lack of background subject knowledge of primary teachers. Initially, topic work became very science-orientated as teachers put a lot of effort into ensuring that this new (for many) area of the curriculum was delivered properly. This was probably an inevitable consequence of the statutory "threat", as well as of the novelty of some of the skills and knowledge to be covered. The pendulum swung in the direction of the humanities in a similar way with the introduction of NC Geography and History. The integrated nature of topic work has been more difficult to maintain at KS2: many schools have opted for a more subject-orientated approach in the upper years of KS2. The Alexander Report, (Alexander, Rose and Woodhead 1992) questioned whether it is realistic to assume that the generalist class teacher, aided by the curriculum co-ordinator, can effectively deliver nine subjects plus Religious Education. The tendency towards specialisation in the upper years of KS2 was predicted as a likely outcome for both teaching and learning. Some support for the science background knowledge of science co-ordinators and classroom teachers was provided, as a result of grant-aid from the DES for "20-day courses". These courses are, in the main, delivered in conjunction with higher education. An Open University distance learning course is also available. The DES courses began in 1989-90 and end in 1993; they are likely to reach only a third of schools. The NCC also embarked, in

1992, on the publication of distance learning materials specifically aimed at enhancing subject expertise in particular aspects of science. Primary school teachers in general feel the need for support with physical science topics such as forces, electricity and magnetism.

1.2.3 Key Stage 3

Science at KS3 had less impact. Science teachers assumed or asserted that they were "doing most of it already", as integrated courses were the norm at this stage. The arrival of the NC gave added impetus to the movement towards balanced science. It was in the areas new to school science, such as earth sciences, astronomy and the history of science, that most activity was stimulated, but in a search for resource materials rather than in any drastic revision of methodology or the structure of the curriculum. Publishers rushed to show how NC compatible their schemes were and produced materials to fill the gaps. Modifications to practice took a little longer in secondary schools than primary schools. Coming to terms with the new emphasis on IT in science was an awareness-raising and in-service training issue, sometimes sidestepped as a problem of under-resourcing – though this is a contributory factor. The recognition of the importance of a comprehensive scheme of work was a salvation for many science departments. Using PoS to define areas of content and enable achievement of attainment targets (which can then be specified in the scheme of work), and supporting this by suggestions for pupil activities, appropriate teaching/learning strategies, assessment and IT opportunities and homework, together help the planning and delivery of NC science. Changes in the management structure of science departments towards supporting and reflecting the organisation of the NC rather than the subject disciplines are also increasingly common, with incentive allowances being identified for responsibility for a key stage, assessment and recording, or organisation of a specific course.

1.2.4 Key Stage 4

KS4 National Curriculum GCSE courses, consistent with NC criteria, started in 1992 for examination in 1994. Double award (formerly model A) and single award (formerly model B) courses are available. Between 1989 and 1992 schools were required to meet the 'reasonable time requirement': that is all pupils had to study science, the majority a double subject equivalent, with a single subject equivalent for the "minority of pupils who might need

".... to spend more time on other subjects, for example to develop a special talent in music or foreign languages" (DES 1989).

The single science option (model 'B') has always been a controversial compromise. Many still feel there is a danger that it could become a "second-class" option for girls or that it will be used for the less able. The reduction from double science is achieved by leaving Sc1 intact and excluding strands from the PoS for Sc 2-4. The continuance of separate science subjects was legislated for by the Secretary of State. These must be taken as a "suite" of subjects, physics, chemistry and biology: existing syllabuses are to be examined in 1994 with new subject criteria for subsequent years. The survival of separate subjects was felt by some, perhaps many, to subvert the broad and balanced science philosophy of the NC.

2 The Structure of the National Curriculum for Science

2.1 The elements that make it up

The National Curriculum is a statutory framework which encompasses what is to be taught and what it is intended will be learnt by pupils or, more specifically, what pupil achievements are to be assessed. It should be remembered that the whole curriculum comprises cross-curricular elements and Religious Education as well as the NC. The "what is to be taught" are the programmes of study (PoS), continuous commentaries explaining what pupils should "do". Programmes of Study are defined as:

> *...the matters, skills and processes which must be taught to pupils during each key stage in order for them to meet the objectives set out in the attainment targets.*

A PoS is, then, more akin to a syllabus than a scheme of work.

The Attainment Targets (ATs) are the learning outcomes and are defined as:

> *...the objectives for each foundation subject, setting out the knowledge, skills and understanding that pupils of different abilities and maturities are expected to develop within each subject area. They are further defined at ten levels of attainment by means of appropriate statements of attainment.*

The NC for science is currently set out in a document called *"Science in the National Curriculum* (DES and Welsh Office 1991) which supersedes the 1988 version. The pages of parliamentary jargon bracketing the document reflect its legal status. The statutory requirement to teach the NC only applies to state schools in England and Wales but there is widespread adoption of it the independent sector. Northern Ireland has it's own NC which, as far as science is concerned, closely resembles that of England and Wales. Scotland has a different approach to the curriculum – it has NO National Curriculum!

(a) Programmes of Study
The Programmes of Study need further interpretation and extension before a teaching programme can be devised which structures and sequences the scheme of work for science in a school. It is possible to group together the guidance in the programmes of study under various headings.

(i)   Those that have a practical emphasis – for example, pupils should ....
    .... be encouraged to develop skills ....
    .... collect and find ....
    .... explore the properties of ....
    .... explore the effect of ....
    .... observe ....
    .... classify ....
    .... have the opportunities to experience ....
    .... investigate practically ....
    .... make predictions ....
    .... have experience of ....
which enable a teacher to identify suitable activities as part of the scheme of work.

(ii) Those that deal more with ideas and concepts e.g. pupils should ....
    .... be encouraged to develop .... understanding ....
    .... learn that ....
    .... be introduced to the idea that ....
    .... be introduced to the concept that ....
    .... appreciate the relationship between ....
    .... explore the nature of .... consider ideas about ....
These will figure as part of the knowledge to be acquired as part of a scheme of work.

(iii) Those where relevant links with everyday life and the impact of science on society are detectable themes. Pupils should ....

.... consider everyday uses of ....

.... develop an awareness of ....

.... relate their experiences to ....

It is useful to use such statements to identify crosscurricular themes in the scheme of work.

(iv) Some imply guidance on teaching method. Pupils should ....

.... be given opportunities to discuss ....

.... to investigate ....

but these are rare, and in general teachers can put flesh on the bones of the programmes of study in their own teaching and learning styles.

The emphasis in the NCC definition of the PoS is on teaching rather than learning, but both DES and NCC have always stressed that the National Curriculum has never prescribed methodology; it states, the "what" but not the "how". This leaves teachers with some freedom to interpret the programmes of study as they would a syllabus. The PoS can be used to initiate planning, or as a check on the scope and coverage of a topic or scheme already planned.

Fig 3 attempts to summarise how the demands of the NC may be incorporated into the planning process for schemes of work, for a piece of science with a fairly small amount of cross-subject linking and integration.

Each PoS is aimed at a Key Stage. Key stages are a way of packaging the years of compulsory schooling. Each key stage ends with national testing at ages, 7, 11, 14 and 16. Key Stage One (KS1) thus covers two years, Key Stage Two (KS2) four years, Key Stage Three (KS3) three years and Key Stage Four (KS4) two years. Programmes of Study are preceded by general statements which set the scene for each key stage and stress the importance of the development of communication skills and other key areas:

– *the importance of science in everyday life at KS1 and KS2,*

– *the application of science and the nature of scientific ideas at KS3,*

– *the application of economic, social and technological implications of science and the nature of scientific ideas at KS4.*

Many of these fundamental themes featured in their own right as attainment targets in earlier versions of NC science. It is interesting to follow through some of the evolutionary pathways in Figures 1 and 2.

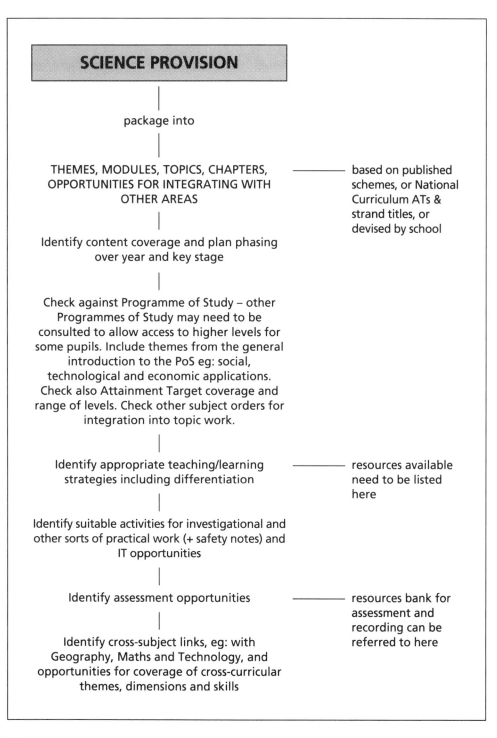

Figure 3 Planning a scheme of work for National Curriculum Science

(b) Attainment targets

Attainment targets are, in effect, learning objectives. They specify what it is expected pupils will have achieved as a result of having been taught the programmes of study. The interpretation of attainment targets and statements of attainment as assessment criteria has superseded the intention of the original authors of the NC science. They regarded them as general indicators of the performance to be expected by pupils at particular levels of achievement. Guidance by means of non-statutory examples helps teachers to visualise the sort of activities that pupils could experience, to enable them to achieve a particular level, in a particular attainment target. The inclusion of examples was a result of the consultation exercise on the revision of the original science national curriculum. There was some reluctance, on the part of the DES and NCC, to include them because of the danger that the examples could become a dogmatic and prescriptive syllabus rather than guidance. Teachers however welcomed them, as ways of interpreting the statements of attainment and providing ideas for classroom activities. The pupil is put very much as the centre of these activities and the teacher is not mentioned ....

pupils could ....
- use information .... from an extended family ... to show that a feature may be inherited ....
- identify processes .... such as feeding, growing ....
- explain how .... the heart acts as a pump in the body
- describe .... a local example of human impact on the environment
- compare .... a candle and a car engine
- identify variables .... which affect the rate at which water cools ....
- sort .... and group materials according to their shape, colour or hardness
- take measurements .... of the biological and physical factors involved at different sites ....
- select techniques .... such as decanting, filtration, dissolving and evaporation
- use knowledge of .... heating, evaporation, condensation .... to explain the water cycle
- deduce .... the products of heating elements ....
- design and make .... a circuit ....
- use diagrams to show .... the direction of forces on objects ....
- calculate .... the mean speed of vehicles from measurements ....
- predict .... the effect of the Sun's mass on the paths of a comet ....

- apply knowledge .... of energy transfers

and of course ....

- investigate!

One weakness of the examples is that they are the same for each level irrespective of the context of that level in a particular key stage. Pupils may be at level 3 attainment in KS1, KS2, KS3 or KS4: a single example cannot be adequately relevant to all these age-groups. This reinforces the categorical (and arbitrary) nature of the ten-level scale but teachers are free to use their professional judgement to provide comparable alternatives. The word 'comparable' however hides an inherent problem in the whole structure: the underlying assumption that what pupils can do in one context they can do in another. Once statements of attainment are used as assessment criteria this becomes a crucial issue: one that has not been fully addressed.

2.2 The science attainment targets

There are four attainment targets for science, *Sc 1, Scientific investigation, Sc 2 Life and living processes, Sc 3 Materials and their properties* and *Sc 4 Physical processes*. Sc1 (or AT1 in references to the first version of National Curriculum Science) is about practical skills, which are to be acquired in the context of the areas of knowledge and understanding specified in the programmes of study for Sc 2, 3, 4.

2.2.1 Sc1

Specific to Sc1 are those skills concerned with the process of scientific investigation, sometimes described as procedural understanding. They come together in the context of a whole investigation, fig 4.

Other kinds of practical work are part of teaching and learning about science but are not Sc1 because they illustrate concepts or processes in a prescribed way or involve practical skills and observational exercises outside the context of an investigation. The different kinds of practical work are summarised in fig 5.

The NC has, in a way, attempted to tease out the essence of what makes science different from other subjects. It has defined science in terms of a body of knowledge and concepts that is peculiar to science and, perhaps more importantly, the way in which science works. The view that it is intellectually impossible to disentangle the two threads and that an artificial division of the two ways of understanding has been made, is probably true, but

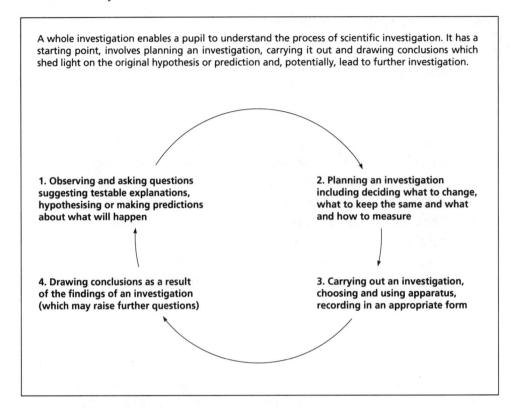

A whole investigation enables a pupil to understand the process of scientific investigation. It has a starting point, involves planning an investigation, carrying it out and drawing conclusions which shed light on the original hypothesis or prediction and, potentially, lead to further investigation.

1. Observing and asking questions suggesting testable explanations, hypothesising or making predictions about what will happen

2. Planning an investigation including deciding what to change, what to keep the same and what and how to measure

4. Drawing conclusions as a result of the findings of an investigation (which may raise further questions)

3. Carrying out an investigation, choosing and using apparatus, recording in an appropriate form

Figure 4 Whole Investigations

the structure and nature of science process is more likely to be appreciated, particularly by non-specialists, with the arrangement that we have.

Giving investigation prominence in the ATs has ensured that this critical aspect has to be tackled by teachers from KS1 upwards. Sc1 has been, and is still, a major challenge for teachers and it is taking them some time to come to terms with it. Investigation for investigation's sake, when content and context are, or appear, trivial, has been criticised. The NC stresses the importance of setting Sc1 in the context of the other attainment targets and drawing upon knowledge and understanding of science, a point constantly reinforced in the Programmes of Study. Sc1 helps pupils to refine and clarify ideas and concepts. An investigation may be used as a vehicle for teaching a particular concept or for reinforcing it by exploring its practical manifestations after pupils have been taught or have learned about it.

What the attainment target structure does is to identify progression in investigational skills in a way which has not been done before and which perhaps does not fit happily within the ten-level framework. Sc1 is too new

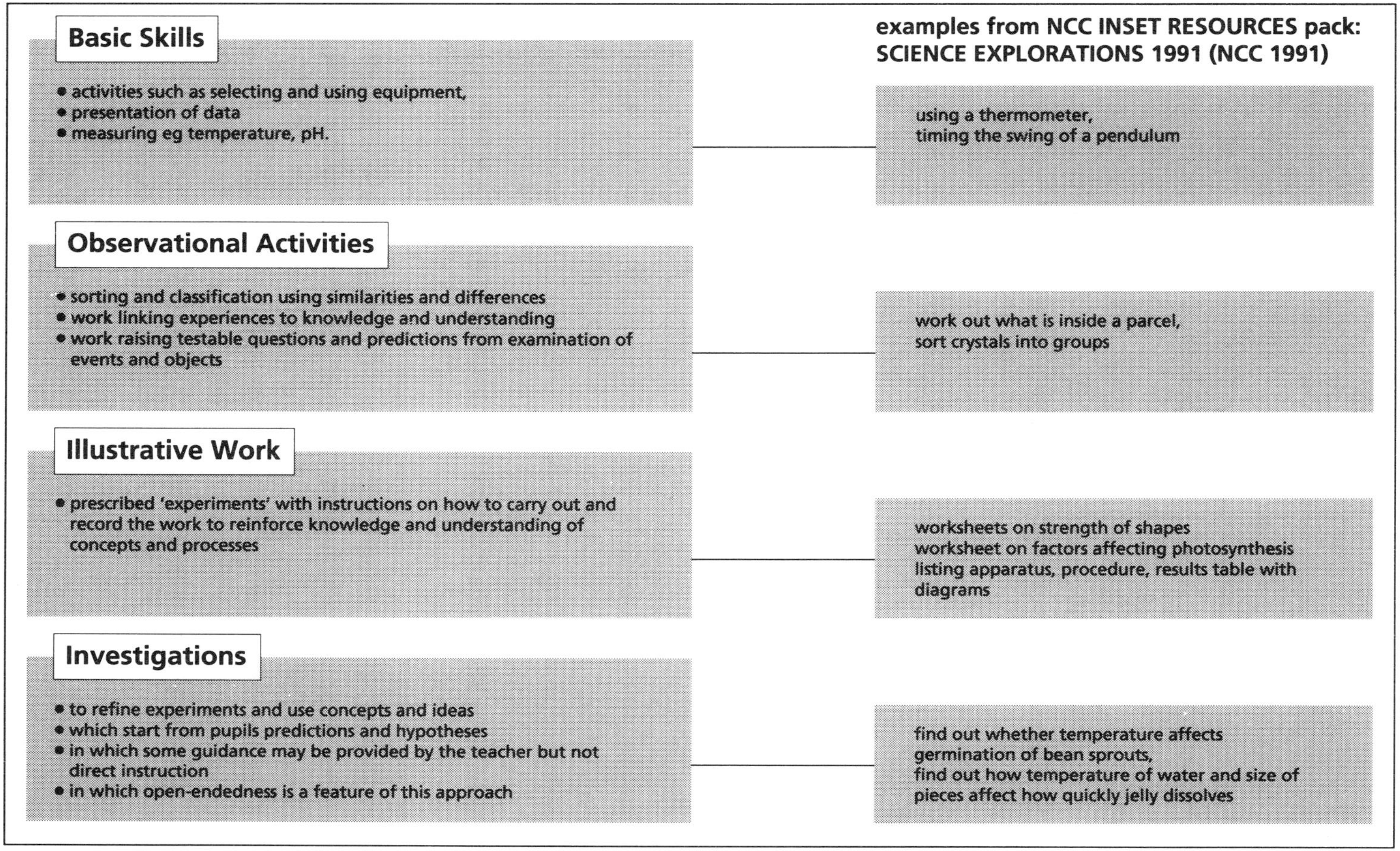

examples from NCC INSET RESOURCES pack: SCIENCE EXPLORATIONS 1991 (NCC 1991)

using a thermometer,
timing the swing of a pendulum

work out what is inside a parcel,
sort crystals into groups

worksheets on strength of shapes
worksheet on factors affecting photosynthesis
listing apparatus, procedure, results table with
diagrams

find out whether temperature affects
germination of bean sprouts,
find out how temperature of water and size of
pieces affect how quickly jelly dissolves

Basic Skills

● activities such as selecting and using equipment,
● presentation of data
● measuring eg temperature, pH.

Observational Activities

● sorting and classification using similarities and differences
● work linking experiences to knowledge and understanding
● work raising testable questions and predictions from examination of
  events and objects

Illustrative Work

● prescribed 'experiments' with instructions on how to carry out and
  record the work to reinforce knowledge and understanding of
  concepts and processes

Investigations

● to refine experiments and use concepts and ideas
● which start from pupils predictions and hypotheses
● in which some guidance may be provided by the teacher but not
  direct instruction
● in which open-endedness is a feature of this approach

Figure 5 The different kinds of practical work in science.

37

for us to know whether it is workable in the reality of the school laboratory. Some of the statements are not easy to interpret in the light of the behaviour of real pupils, and it has yet to be seen if the resource implications of open ended investigational work can realistically be met.

The identification of progression in scientific investigation is not easy, particularly when it is required to fit the specific framework of an attainment target. Proficiency in investigation is probably only achieved by doing investigations: the idea that it can somehow only be acquired after practising skills, observational activities and illustrative practical work is liable to delay the acquisition of investigational skills. It is clear that "learning how to investigate" is an area where materials need to be developed.

The components of investigational (or procedural) understanding in Sc 1 are summarised in terms of three strands:

(i)   ask questions, predict and hypothesise;
(ii) observe, measure and manipulate variables;
(iii) interpret results and evaluate scientific evidence.

These are not to be seen as separate skills but as complementary aspect of whole investigations. Pupils may progress at different rates in these three aspects but all three must be developed together.

"Stranding" Sc1 in this way has produced a more accessible approach to investigations which has many similarities with the "Planning, Implementing and Concluding" version of science process adopted by the Graded Assessment in Science Project (ULEAC 1992). For recording progress and to help diagnose where pupils need help to improve, teachers may assess one, two or three strands from a single investigation. When a level has to be reported at the end of a key stage, teachers will be able to aggregate the three strand levels using their professional judgement rather than a formula. If a pupil scores level 1 for strand (i), 2 for strand (ii) and 3 for strand (iii), the teacher may choose an overall level 2 as an average, level 3 using the 'best' level or level 1 the "trailing edge" level! At the time of writing it is not clear how the teacher assessment of Sc1 will be moderated by any quality audit process. Fig 6 attempts to summarize Sc 1 by picking out the main features at each level of the attainment target.

2.2.2 Sc 2, 3, 4

The Attainment Targets Sc 2, 3 and 4 hark back to the first attempts of the science working group in 1987 to categorise science content. That, in turn,

	Plan and carry out investigations drawing on an increasing understanding of science		
Strand **Level**	**(i)** **Hypothesising & Predicting**	**(ii)** **Carrying Out**	**(iii)** **Interpretation & Evaluation**
Level 1	Observational		
Level 2	Predictions & observations		Compared
Level 3	Predictions	Tested	Fairly?
Level 4	Predictions based on prior knowledge	Fair test with selection of equipment and measurement	Draw conclusions
Level 5	Hypothesis Construction	Range of variables chosen to test hypothesis	Conclusions evaluated
Level 6	Predict and investigate the relationships between continuous variables		Explain relationship and results theoretically
Level 7	Predict relative effect of two or more independent variables	Compare effects	Draw conclusions on relative effects and limitations of evidence
Level 8	Make quantitative predictions based on theory and decide what to investigate	Select instruments to suitable accuracy	Evaluate effectiveness of strategies in determining relative effects of individual factors
Level 9	Use a theory to make quantitative predictions. Collect valid data	Systematically using a range of techniques to judge relative effect of factors	Analyse in terms of complex functions, calculations, graphs, etc.
Level 10	Hypotheses based on laws, theories and models	Collect data to enable	Evaluation of a law, theory or model as an explanation of observed behaviour

1. Identification and control of variables begins at level 4 as part of 'fair testing'
2. Evaluation begins as part of the 'fair testing' process

Figure 6 Summary of Sc 1: Scientific Investigation

had been based on *"Science 5-16: A Statement of Policy"* (DES 1985 and Welsh Office). Sc2 packages the biological and ecological aspect of science. The physical sciences are split between Sc3 and Sc4, Sc3 being mainly chemical with earth science and Sc4 being physics with astronomy. The individual strands reflect their pedigree through the various stages of the science NC. Many strands were once attainment targets in their own right: fig 2 (p 27) attempts to trace the ebb and flow.

3 The Revision of National Curriculum Science

3.1 The rationale

The reduction in the number of attainment targets and the abandonment of profile components was the result of a ministerial decision, announced in January 1991, that the curriculum needed simplifying, to make assessment more manageable. The revision was carried out by HMI in consultation with NCC. They used the opportunity to improve progression and continuity in the attainment targets, using research into children's learning in science which had taken place since the original version. This creativity was strictly outside their brief, as a major review was not intended, merely "re-brigading" – to use the jargon of the time!

The proposals for revision were initially met with more resistance than the revisers expected. Teachers had taken considerable trouble to come to terms with the original targets. They felt that the hard work that they had put in had been wasted and that the rules had been changed yet again.

3.2 The changes

In the event, the number of statements of attainment was reduced from 409 to 176. Some statements changed levels and in general there is now a greater consistency in their scope, though some are still much broader than others. Some of the "pruned" Statements of Attainment (SoA) appear in the Programmes of Study but are still liable to be tested. Reference to fig 2 illustrates the expansion and contraction of the National Curriculum over the years. One cannot help reflecting that in some ways the wheel – if it has not been reinvented – has come full circle. At least the evolution of NC Science suggests a combination of natural and artificial selection with the final version looking more manageable. Fig 7 shows the origins of the strands, and is the sort of analysis done when converting schemes of work from the 1989 to the 1991 orders.

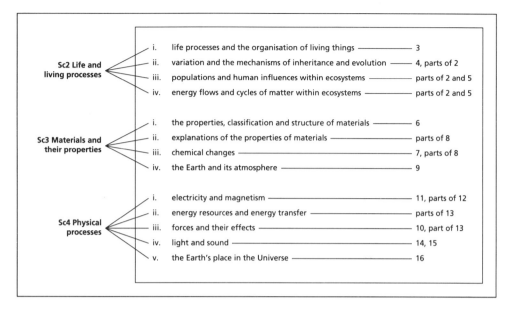

Figure 7 Origin of Strands Sc 2, 3, 4

3.3 Assessment influences

The 1989 orders, though demanding, gave much more detailed guidance than the current version on what was to be assessed. Primary teachers, in particular, find some of the new SoA too generalised and still rely on the 1989 orders for interpretation and amplification. The streamlining process has resulted in a reduction in emphasis on social implications, the history of science, multicultural aspects, health education and communication. At least that is what transference from SoA to PoS may cause, even though PoS are statutory. Aspects not assessed may well be de-valued in a curriculum that started life as a basic entitlement but has become all-embracing.

The issue of how trends in national assessment will affect teaching is a real one. The prospect of paper and pencil tests at all key stages, rather than the extended assessment tasks originally proposed by TGAT is bound to make teachers less willing to stray far from the documentation. The balance between national tests and teacher assessment has also undergone various modifications. Originally it was suggested that some kind of combination of the two, or moderation of one by the other, would result in a final level for the subject being determined. In practice, the reality is of short national tests for assessment of knowledge and understanding (conceptual understanding) and teacher assessment for Sc1 (procedural understanding). There is evidence from KS1 assessment that, in the absence of a standard task for Sc1 in 1992, teachers have devoted less time to it, and that performance by pupils in Sc1, as measured by teacher assessment, has gone down. There is a danger that a vital vehicle for learning in a fundamental aspect of science could become neglected.

The nature of the KS3 assessment – 3 × 1 hour pencil and paper tests for Sc2, 3, 4 and teacher assessment for Sc1 – may influence practice in this key stage. The testing of three years work at the end of Year 9 and four years work at the end of KS 2 certainly calls for some kind of strategy to cope with it! The most acceptable answer seems to be that a revisiting of attainment targets and strands is planned for, with revision in the context of current, related, learning. A regime of revision tests punctuating the KS3 or KS4 course at regular intervals is a depressing prospect. Pupils' written work will be kept for reference and their organisational and study skills will need more practice and thought than in the past.

At KS 4 assessment of Sc1 will replace current schemes of assessment of practical skills. The requirement for assessment of investigational skills in the context of a whole investigation will require some re-alignment of existing practical assessment. Departments will differ in their capacity to integrate Sc1 into the normal run of work. In practice, continuous assessment of practical skills in many schools has turned into a series of 'set-piece' practical exams: the difficulty of assessing an individual's contribution to group work has tended to reinforce this strategy.

The complexity of provision at 14+ extends beyond the issue of which GCSE course to choose, though even that is difficult enough. Potentially, the introduction of vocational courses can increase the repertoire of a science department. The uncertificated levels of GCSE are also generating courses. How the key stage will evolve in the light of the Secretaries of States wish for "more flexibility" remains to be seen. Once again, decisions taken nationally to change the rules have to be sorted out in detail afterwards by teachers.

4 Conclusion

Both primary and secondary teachers have shown remarkable resilience and stamina in the way in which they have tackled the nine versions (yes nine – count them on fig 1 (p 26) of the NC Science that they have had to consider or implement since 1987. As a period of some stability is expected for science before the next review, it seems likely that schools will have an opportunity to refine existing practice and even consolidate in the light of experience. Whether there will be time for reflection is doubtful. The quality of the guidance offered by the National Curriculum is high. It was always intended to reflect good practice; what has been missing has been the time to plan for change. Changes have been forced on schools as a result of policy and logic beyond the control of teachers. The National Curriculum has been described as "a cure for which there is no known disease" by Professor John Tomlinson (quoted in Times Educational Supplement 1992). The measure of its success will be whether the quality of education after its implementation is better than before it. The reaction in science is probably "we were getting there quite nicely on our own thank you!". How OATs (Old Attainment Targets) became NATs (New Attainment Targets) after briefly being PRATs (Proposed Revised Attainment Targets) is now history, but you may still hear more experienced (older) colleagues musing about GOATs (Good Old Attainment Targets!).

David Oakley is General Inspector (with responsibility for Science) for Dudley LEA. He taught science for over 20 years in comprehensive schools, a tertiary college and adult education, before becoming a Science Advisory Teacher. He was Chair of the Science Advisers and Inspectors Group of the ASE 1991-92.

References

Alexander R, Rose J, Woodhead C. (1992) *Curriculum organisation and classroom practice in primary schools: a discussion paper*, DES.

Department of Education and Science and the Welsh Office (1985) *Science 5–16: A Statement of Policy*, HMSO.

Department of Education and Science (1989) Circular No.6/89 The Education Reform Act 1988: National Curriculum: *Mathematics and Science Orders under section 4*, HMSO.

Department of Education and Science and the Welsh Office (1991) *Science in the National Curriculum (1991)* HMSO.

ULEAC (1992) *GCSE Science Syllabus: Graded Assessments in Science Project (GASP)*, ULEAC

INSET Resources pack: *Science Explorations (1991)*, NCC.

Teaching and learning science

3 Learning Science

Hilary Asoko, John Leach and Phil Scott

1 Introduction

'..so what has learning to do with teaching...?'

At first sight this seems a ridiculous question. Learning follows from teaching; teachers teach and pupils learn. At least that is the theory of it. However, as anybody who has ever taught knows, things don't always work out like that! Experienced teachers are well aware of those parts of the science curriculum which create major problems for all but the brightest of pupils. Why should that be? How might teaching approaches be evaluated and modified to help pupils overcome those difficulties?

Perhaps one answer is to give more attention to what is involved in learning science. Teaching methods whether they be 'process based', 'individualised programme' or 'discovery' approaches, should be informed by views of learning. Although in the real world of the classroom other factors also need to be taken into account, an understanding of how science is learned has an important part to play in deciding how best to teach science. It is no accident that in this handbook the chapter on 'Learning' precedes the chapter on 'Teaching'.

2 What do we Know about How Children Learn Science?

Learning science involves a number of distinct components:

- acquiring science concepts
- developing science skills and processes
- appreciating the nature of science and the role of science in society

This section examines what is known about how children learn these three components of science. Developmental influences on learning and the importance of language and context are also considered.

The view of learning adopted in this chapter is broadly constructivist in nature. Learners are seen as active and intelligent. They bring a range of

46

ideas, attitudes, abilities and skills to any learning situation. The learning process involves an interaction between the learner and any learning experiences presented, whether these are observations in science classes (or elsewhere), or explanations from teachers or others. Learning involves the learner in making sense of things in terms of their existing ideas, though this will sometimes involve moving beyond their current interpretive framework to one which is better able to make sense of their experiences, including the recent experiences deliberately provided by their science teachers.

2.1 Children's conceptual understanding in science

By the time children start school, they have already had considerable experience of natural phenomena and events in the world around them – light and shadow, heat and cold, falling objects, floating objects and so on – and have begun to develop ideas and explanations about these. Personal observations and experiences play a part in this development as do conversations with other children and with adults and influences from television and from books. It is these which children must draw upon, in science lessons as well as everyday life, when making sense of new experiences, when making predictions or observations and when trying to understand new ideas.

In general, since pupils come to science lessons with broadly similar backgrounds and experience, it is not surprising that the ideas they bring are not entirely idiosyncratic but tend to fall into patterns which change according to the age and experience of the child. Research from around the world has provided information about the ideas and understandings which children characteristically develop in a number of topic areas. For example many 11–16 year olds think that:

- light travels further at night than in daytime
- matter disappears when substances dissolve or burn
- vacuums 'suck'
- plants take in food through their roots
- the temperature of an object depends on the kind of stuff it is made of (e.g. metals are cold, plastics are warm)
- air is weightless or has negative weight

Sources of information about children's ideas in science are listed at the end of this chapter.

As children gain more practical and social experience of the world around them, their explanations for certain phenomena may change. Figure 1 shows the responses of children aged from 7 to 16 years when asked to consider a balance, with containers of water and sugar on each side. They were asked what would happen if the sugar on the left side of the balance was tipped into the water and the containers replaced in their original positions. The graph shows the percentages of children who predicted that the balance would remain level, together with those who anticipated that the left side would gain or lose mass. (Johnston 1990 p 25)

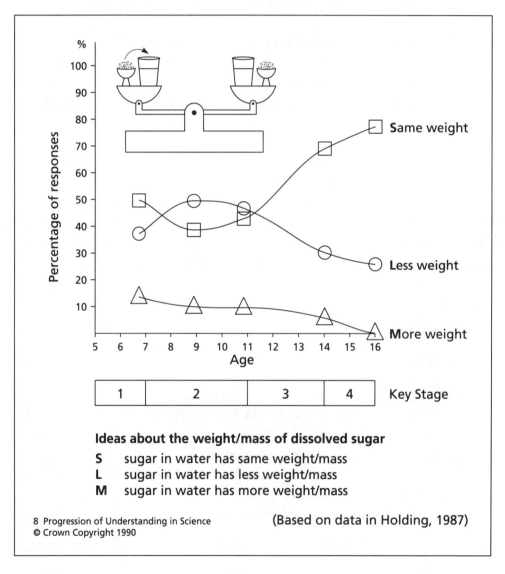

Ideas about the weight/mass of dissolved sugar

S   sugar in water has same weight/mass
L   sugar in water has less weight/mass
M   sugar in water has more weight/mass

8 Progression of Understanding in Science
© Crown Copyright 1990

(Based on data in Holding, 1987)

Figure 1 Prevalence of ideas about the weight/mass of sugar that has disappeared in water (n = 588)

In making predictions children draw upon their ideas about dissolving together with their notions of weight. Children who predict that there will be less mass on the left hand side may do so for a variety of reasons. Younger children may believe that the sugar has 'just gone' and thus can have no weight. Once they recognise that the sugar is still there, but invisible, they start to consider what happens to it when it dissolves. Many, using the idea that the sugar breaks up into 'small bits', predict that the bits are so small they weigh nothing, or that because they are 'floating' they weigh less than normal. Children who focus on the sugar and see it as 'soaking up' water may conclude that it becomes heavier and thus predict a gain in weight. By age 13 onwards, pupils are more likely to consider the sugar as an amount which does not change, regardless of form.

When children learn science they are often asked to consider familiar events and phenomena in new ways. For example, when the action of drinking through a straw is explained in terms of differential air pressures, it would not be realistic for the teacher to assume that the pupils do not already possess views on such matters. Of course they do, and science teachers will be familiar with pupil explanations in terms of 'me sucking' or 'the air inside the straw sucking' or 'the vacuum in the straw sucking'. In this particular example the science explanation is quite contrary to informal views and is based upon the action of external pressure whereas the informal views are based upon an internal sucking force. Learning the science view in this context must therefore involve a significant restructuring of ideas by the pupil as he or she comes to understand a mode of explanation which initially may be counter to 'common sense'. It is important to note that scientific ideas such as air pressure, atomic structure or photosynthesis cannot be discovered by direct observation but must be constructed as part of complex and abstract explanatory models.

Children's conceptual understanding of science appears, then, to develop when concepts are exchanged, modified or refined. The development of ideas in one topic area may be dependent upon the progression of ideas in a related area. In general terms, children move from models which are based on easily observable features and which relate to a limited number of phenomena to those which depend on abstract entities and which have more general use.

The National Curriculum provides a framework that aims to lead pupils to increasingly more 'difficult' concepts, but the progression in children's understanding does not necessarily develop in the additive and linear way suggested.

2.2 Children's procedural competence in science

Practical work holds a central place in the teaching of science in the United Kingdom. Ideas about the nature and value of practical work have often been debated, but, since the 1960's, major curriculum developments have encouraged the use of hands-on practical activity. The intended purposes of such activity include: development of scientific skills, illustration and development of scientific concepts and provision of opportunities for scientific investigation. In addition, practical work is often seen as being important in motivating pupils and developing attitudes such as curiosity, objectivity and willingness to evaluate evidence.

Pupils bring to their practical work, as to any other aspect of their science learning, ideas and expectations. Children's procedural competence will be influenced by their conceptual understanding, their views about the nature of the activity they are involved in and the context in which the exercise is set. With skills such as measurement, for example, a child's performance may be affected by the type of scale used, whether digital or analogue, their understanding of the divisions of the scale and perceptions of the level of accuracy needed. A pupil who can use a thermometer competently in a structured situation may find difficulty in deploying this skill in an investigation where decisions need to be made about the number of measurements to be taken and when to make them.

Classification provides a further example. Using the characteristic properties of metals and non-metals to classify materials requires a very different conceptual background from that needed to classify arthropods into subgroups. While both tasks may require observation, it is not simply a matter of 'seeing' but also of knowing what to look for, and how to find it.

Practical activities are often used to introduce pupils to science concepts. This can, however, be problematic. In the dissolving example detailed in the previous section it might be thought that pupils could be convinced by a simple demonstration that mass is conserved. Unfortunately, pupils do not always 'see' what they are intended to see, and even minute movements of the balance pointer may be taken as evidence that supports their view and refutes the science idea. Unless pupils have a good grasp of what they are looking for and why, and what the possible outcomes mean, they are unlikely to be convinced by 'critical experiments'.

The situation is even more demanding in investigational work. In order to plan and carry out a successful investigation, pupils must be able to bring together conceptual understanding with process skills such as measurement, hypothesising, predicting, observation, control of variables and recording and interpreting data. There are thus both procedural and conceptual demands to any investigation and these inevitably interact. For example in order to manipulate variables pupils must identify those which are relevant before deciding which should be systematically varied, which are the dependent variables and which must be controlled.

Models of progression in investigations have been suggested in Sc1 of the National Curriculum. These include shifts from single to multiple variables, from discontinuous to continuous variables and from qualitative to quantitative solutions which include derived variables. In addition there are shifts from investigations drawing upon concrete experiences to those involving abstract concepts and from everyday to novel contexts. Whilst such models may provide some way of estimating the procedural demands of an investigation, they ignore the conceptual demands which will influence a child's performance. Further information on children's procedural competence and the factors which influence performance can be found in publications listed at the end of this chapter.

2.3 Children's ideas about the nature of science and the role of science in society

Our society presents many images and facets of science and it should therefore come as no surprise to find that pupils have their own ideas about the nature of science - what science is all about, what experiments are for, and so on. Science teachers bring certain assumptions about what science is to their teaching, and most of these assumptions are likely to be shared by other scientifically literate adults, including professional scientists. Readers interested in the philosophy of science are directed to Chalmers 1982. The following paragraphs describe children's ideas about the purposes of science, and the relationship of scientific theories and evidence. This is set alongside a brief description of the sorts of assumptions about the nature of science that underpin science teaching, and possible implications for learning science are suggested.

What do we know about children's ideas about the purposes of science? It is difficult to find one purpose for the work done by very different scientists such as astronomers, industrial chemists and clinical pathologists. Some of the work involves creating new knowledge, but in other cases it involves using existing knowledge to create new materials, develop new techniques or perform routine tests. In general, however, the purposes of science can be summarised as explaining and controlling the natural world.

What do secondary school pupils see as the purposes of science? In some cases, pupils do not think that science is involved with explanations, but rather that the purpose of science is to make things happen. Teachers may well remember lessons on photosynthesis in which pupils have tested leaves for starch and interpreted their results as having 'worked' for a plant in the light, but 'not worked' for a plant kept in the dark. For these pupils, the purpose of the experiment is to show that leaves may be stained blue-black with iodine solution, rather than to consider a theory that plants make starch in the presence of sunlight.

Some pupils may think of science as an investigative process, but in a mechanistic way. For example, they may think that scientists find out whether different fabrics are waterproof, rather than using their knowledge of why fabrics are waterproof to devise new and better ones. For these pupils, investigations involve discovering facts, and explanations do not feature in their reasoning.

It is very common for young people to think that the purpose of science is to help people by creating devices, medicines and so on of benefit to human beings. They may not, however, appreciate that this involves the application of theoretical knowledge to particular problems.

When pupils approach activities such as investigations in school it is assumed that they are able to evaluate theories against the evidence, and that they see the purpose of the investigations as being to do with explaining things rather than making something happen. As can be seen, this may not be the case.

What do we know about children's ideas about theory and evidence? Although there is by no means complete agreement about the nature of theories and their relationship to evidence, a number of assumptions are widely accepted, for example that scientific theories are conjectural in nature, being developed by creative, imaginative processes in the mind. Furthermore, scientific theories should be generalisable to as diverse a range of phenomena as possible, and they should not contradict each other.

Secondary school pupils often have very different ideas about theories and evidence from those described. In most cases, pupils are likely to think of scientific theories as 'facts' which scientists 'discover' by careful observation. For such pupils there is no element of creativity or imagination involved in the interpretation of data. Because they do not distinguish between evidence and theory, some pupils may not be able to use observations to evaluate a particular theory.

Secondary pupils may not think of theories as having to explain as many different phenomena as possible. Science teachers will be familiar with pupils who use different explanations in different contexts, and do not seem to see any need for one common explanation.

The following discussion illustrates this point. A researcher is interviewing pupils after a lesson on plant nutrition, in which the teacher had presented views on photosynthesis which contrasted with the views of most pupils about plant nutrition:

Interviewer: Most people in this class put that plants get their food from the soil and Ms. L*** said that plants DON'T get their food from the soil! What have you got to say about that? Does that surprise you?
Pupil 1: No..
Pupil 2: It surprised me cos I always thought that plants get their food from the soil and all the goodness in it. And when it's been raining the goodness goes into the soil and they get that...
Interviewer: So what she's saying and what you thought are different?
Pupil 2: Well we don't have to agree..
Pupil 1: Well, none of us has to be wrong do we cos we (pause) cos they could do both (pause) to grow..

As children's ideas about the interplay between theories and evidence become more sophisticated, it is likely that their learning of science concepts through a range of teaching approaches will be improved.

Within the scientific community, theories come to be accepted by a process of deliberation and judgment. The decisions made by communities of scientists are, however, affected by many factors other than their observations of the natural world! The interests and values of the scientific community, and of society at large, influences the work that scientists do, the decisions they make, and what counts as being science.

In general, pupils seem to think that theories come to be accepted as a result of being 'true' or 'false', and that competent scientists would all reach the same conclusions from a given set of data. In school, the teacher acts as a kind of 'final arbiter' in evaluating pupils' explanations. For scientists working at the frontiers of knowledge, of course, there is no equivalent process. It is often argued that investigations set in school can act as a role model for pupils of the work of professional scientists, though it is clear that such investigations have a limited capacity to represent either the methods of working or the diverse interests of scientists.

Later in this book, Robin Millar discusses how ideas about the nature of science, and knowledge of science as an enterprise, might be encompassed in adult scientific literacy.

2.4 Developmental processes and children's learning in science

By 'developmental processes' we mean processes that happen as part of maturation, irrespective of teaching. It may be that certain processes already discussed, such as the ability to relate theory and evidence, are developmental. Alternatively, it may be that without practice or teaching they would not develop at all, or would develop more slowly. Are there more general developmental patterns in learning that apply across a broad range of areas of science? Knowledge about these patterns, if they exist and could be documented, would be very useful to science educators!

The work of the Swiss genetic epistemologist Jean Piaget, carried out in Geneva from the 1920's until his death in the 1980's, has been very influential in shaping science education around the world. Piaget viewed learning as an interaction between the existing knowledge structures of learners and their observations of the natural world, resulting in changes to the individual's knowledge structures. He documented children's ideas at different ages about a range of phenomena, and age-related patterns in learning were suggested that were applicable to children's learning across different domains.

In brief, it was suggested that children's reasoning could be characterised according to their ability to, for example, classify, recognise simple cause-effect relationships or handle variables. Young children's thinking was described as concrete whereas the thinking of older children was described as formal, and movement from concrete reasoning to formal reasoning was viewed as a developmental process.

This theory of learning, involving general developmental stages, has been applied to the science curriculum in various ways. Particular activities have been analyzed for their cognitive demands (concrete reasoning or formal reasoning) and placed in the curriculum according to whether pupils are likely to have the appropriate reasoning skills at a given age. The Nuffield 5-13 course is based on such analysis.

The validity of general developmental stages in thinking has, however, been questioned. There is a problem of classifying particular pupils' reasoning into different stages in that many pupils perform differently on tasks in different contexts. Pupils who conserve matter in the case of a solid (such as a lump of plasticine being divided) may not do so in the case of a liquid (such as water being poured between different containers). The stage model does not take into account the particular conceptual demands of different science topics.

A further question concerns the issue of whether teaching can influence pupils in moving between stages. Does the teacher wait for the learner to mature to formal thinking before presenting activities and ideas requiring 'formal' reasoning? Will the pupil then be capable of formal thinking across a number of different contexts?

The Cognitive Acceleration in Science Education (CASE) project has published materials aimed at developing the general reasoning skills of pupils in science contexts (see 'Further Reading' for reference). The CASE project's perspective on learning is only partly developmental in that, while accepting that learners have cognitive competencies similar to those described by Piaget, it claims that the development of formal reasoning can be speeded up by teaching. The cognitive level of pupils is seen as a limiting factor in what pupils can achieve, but it is acknowledged that there are additional conceptual challenges posed by particular areas of subject matter.

In addition to developmental changes which are independent of particular concept areas, Piaget and his co-workers mention developmental changes in specific concepts such as the development of an atomic model of matter. A view of development in particular concept domains is discussed by Carey who writes:

> *I am not arguing that there are not domain-general changes in the developing child's conceptual system ... Rather, I am arguing that much of the evidence that has been taken to support such changes actually reflects domain-specific structural reorganizations. These can, and should, be studied in their own right, for only after they are understood will we be able to sort out the issue of the nature of domain-general constraints on the child's cognitive structures.*
> (Carey 1985 p 192)

Children's developing understanding of the nature of science is not, of course, wholly independent of their learning the concepts and processes of science. Many of the questions about how exactly they are related remain, however, unanswered. We know, for example, that many children do not use all the evidence when they evaluate theories, but is this because they have not yet developed the ability to do so or because they don't have the necessary concepts to see how the theory and evidence relate? Does the science they do in school influence pupils' ideas about the nature of science, and if so, how? If pupils are able to relate theory to evidence in one context, can they do it in another? Do pupils' ideas about the nature of science relate to their ability to learn science concepts, particularly through investigations?

2.5 The importance of language and context for learning.

There are other, more general, influences which may affect children's learning of science. Examples of such influences are language and teaching context.

Although science provides a new way of looking at the world, it is the same world that children experience every day, and is often described using the same words. Knowing when, and how, to deploy scientific meanings and explanations, and distinguishing between these and their everyday equivalents have to be learnt. Difficulties can arise, or be compounded, if the differences between the everyday and the science use of language are not recognised. This might occur in a number of different ways. For example:

- many words in everyday usage also have a scientific meaning and these do not always agree. To a biologist the term 'animal' refers to any living thing which is not a plant; in everyday talk 'animal' tends to be synonymous with 'mammal', thus birds and insects are not usually referred to as animals.
- words which have specific meanings in science may be used in an undifferentiated way in everyday language, examples being force, energy and power; dissolve and melt; and mass and weight.
- metaphors and expressions common in everyday talk may suggest ideas which conflict with the science view point. A model which explains how we see in terms of rays emanating from the eyes is supported by expressions such as 'looking daggers' or 'sending dirty looks'. It is not surprising that a child believes plants get food from the soil if father 'feeds' his tomatoes.

The contexts within which work is set can cause confusion for learners. One can try to raise the relevance of a subject by relating it to, for example, industry, pupils' interests and hobbies, or social and environmental issues. Many such settings are highly complex and require a balance between scientific thinking and, for example, economic or social considerations, so pupils are faced with the additional problem of recognising when and how to apply their science knowledge.

One aspect of the relevance issue which has tended to be overlooked can be invoked simply by asking the learners for their point of view. Once learners recognise that they do have thoughts about the matter in hand the question begins to assume some personal relevance which can provide the necessary impetus for them to pursue it further.

3 Summary and Implications

Learning science involves developing a new way of looking at the world. The initial ideas, attitudes and competencies which pupils bring to science lessons may not, of course, match with the science perspective and we can gain some insight into the demands of learning particular topics in science by comparing pupils' starting points with planned teaching goals. In some cases there may be common ground and learning may involve a development of existing ideas and approaches either in terms of depth of understanding or breadth of application. For example pupils at Key Stage 2 have few problems in identifying the vibration of a guitar string as being responsible for sound production. The challenge for them and for older pupils is being able to extend the 'vibrations idea' to all sound producers, including ones where vibrations are not obvious.

In other areas, pupils may need to consider ideas or experiences with which they are unfamiliar and make connections to existing thoughts.

e.g. At Key Stage 3 most pupils will know that water is needed for rusting to occur, very few will recognise that air (oxygen) is also an essential factor. Learning in this case involves taking on a new idea to extend an existing model.

There are also areas where the science view seems entirely contrary to pupil's existing ideas and considerable effort will be required to develop a different model and to evaluate its usefulness.

e.g. At Key Stage 3 most pupils will assume that a steady overall force is needed to maintain steady motion. This view is quite contrary to Newton's First Law and so learning in this field involves a considerable intellectual 'about turn' for pupils.

In each of these examples learning involves more than 'just' coming to understand the science view. Pupils also need to develop an appreciation of the power of these ideas in terms of their ability to explain a wide range of phenomena. As we have seen, pupils do not necessarily start their science lessons with a commitment to generally applicable ideas, so here we have an additional kind of learning demand.

We are led to an interesting question about one of the fundamental aims of science education. Should we be aiming to replace our pupils' existing ideas with the science point of view? Many would argue that such an approach is neither desirable nor possible. Perhaps a more realistic aim is to help our pupils towards an understanding of the science perspective and, crucially, to help them appreciate the contexts in which it is appropriate to use either scientific or other forms of thinking.

4 Matching Teaching to Learning

Given the preceding views about learning science, what can we say about teaching?

4.1 Start where the child is at

"Ascertain what the learner already knows and teach accordingly.'

Many will be familiar with Ausubel's dictum and it would be fair to say that the need to 'start where the child is at' has become part of the accepted rhetoric of science education. An issue which is perhaps more thought provoking surrounds the way in which such views might actually inform and guide teaching. What is involved in 'teaching accordingly' and how can that be informed by insights into 'what the learner already knows'?

One way into this problem is to return to considering the nature of starting points, teaching goals and consequent intellectual demands. At the onset of planning a particular topic the teacher might usefully ask the following questions:

1. What existing ideas and understandings, relating to this topic, are the pupils likely to have?
2. What are the intended learning outcomes for the topic?
3. What changes must pupils make in their thinking if they are to start with existing ideas and come to terms with the science view?

By thinking through questions such as these the teacher is making an explicit effort to consider the learning demands of what is to be taught and is thereby in a more informed position to plan and implement teaching activities which will support learning. Consider now two brief examples of such thinking in action.

4.2 Two examples of teaching planned to 'start from where the child is at'

Dissolving and melting

A year 7 class were working on 'dissolving' and the teacher noticed that many of the pupils used the terms 'melting' and 'dissolving' interchangeably, which is common in everyday life. The teacher recognised this as being a problem and decided to find out the extent of the confusion by setting the pupils a card sort activity. Written on the cards were statements such as:

Soap .... in bath water ....
An extra strong mint .... in the mouth
An ice cube .... on the kitchen table.

The pupils, working in pairs, sorted the set of 20 cards into two piles, 'Melts' and 'Dissolves'. The teacher then collected the responses from each pair on to a table on the board. By going through this table of information and asking pupils to justify their choices the teacher was able to bring out into the open some of the pupil thinking in this area and to introduce the science distinction between melting and dissolving.

Air pressure

The teacher of a Year 8 class was preparing a unit of work on air pressure. She was aware, both from prior teaching and from reading about children's alternative conceptions, that her pupils' views about the action of items such as 'syringes' and 'kitchen plungers' would almost certainly be in terms of an 'internal sucking force' (e.g. the air inside the syringe sucks up the water). She was also aware that such informal ideas are quite contrary to the science perspective which involves the action of external air pressure.

The teacher reasoned that the pupils' existing ideas would not provide a fruitful starting point for developing the science view (recognising the polar differences between them) and therefore decided to plan her lessons around a series of practical activities and demonstrations designed to help her pupils construct and practise the air pressure view. At the end of the teaching each member of the class had to explain, in terms of air pressure, three simple phenomena. The same questions were given to the whole of the school's teaching staff (!) and the class really enjoyed the final lesson during which they classified the staff responses in terms of 'Old Way' (informal ideas), and 'New Way' (science ideas). This final activity gave the pupils an opportunity explicitly to compare informal and scientific ways of looking at the same phenomena and helped them make connections between the two. It also gave them the chance to reflect upon their own learning.

4.3 Different topics, different approaches

It is interesting, and instructive, to compare the ways in which each teacher in the previous examples 'started where the child is at'. In the first case the teacher used a simple (and enjoyable) activity to collect information about his pupils' thinking. This information was then directly used, in class, as a basis for introducing the science distinction between melting and dissolving. In the air pressure example the teacher was aware of the ideas her pupils were likely to have but did not directly attempt to elicit those ideas in class. In this case 'starting with the child' involved starting with insights to children's thinking and using that as a basis for planning teaching, rather than asking directly what they thought.

The fact of the matter is, of course, that 'starting where the child is at' is fundamentally a statement derived from a view of learning. It can offer a guiding direction for planning and implementing teaching but cannot, in itself, provide an algorithm for teaching. The teacher is still presented with choices and decisions relating to selection of teaching approaches and activities.

In both cases, talk and discussion featured prominently in classroom activities. Many would argue that it is largely through talking about events and ideas that we are able to form and develop concepts. The work of the Soviet psychologist L.S. Vygotsky, carried out during the 1930's, characterised learning as a process of 'enculturation' in which the role of language is central to the child's learning of the values and knowledge of a culture. If science is seen as offering a new way of looking at the world then pupils need time and space to talk about, and practice, their developing ideas. In some situations this might involve pupils in discussing their own thoughts, in pairs or in small groups; at other times pupils might be asked to present the scientific explanation for a phenomenon to the rest of the class or to the teacher. Whatever the context, pupils should be encouraged to articulate their ideas, and if they are to do so, teacher and pupils need to work together to establish a receptive and non-threatening atmosphere.

A final point concerns the issue of 'active learning'. In the two classes, when were the pupils involved in 'active learning'? When they were:

doing the card sort?
listening to other pupils' views?
listening to the teacher?
carrying out practical activities?
classifying responses to questions?

The answer must be that for any single pupil, all, some or none of these might involve 'active learning' - it is impossible to imagine any learning that is 'inactive'. The point is that we cannot equate 'doing science' with 'learning science'. The teacher who is attempting to match teaching to learning will draw upon a range of activities and make selections with particular purposes in mind. Each of these activities has the potential to support 'active learning', it is the pupils who act as final arbiters in determining whether or not they do so.

All three authors are in the Children's Learning in Science Research Group at Leeds University and all joined this group having taught in schools. Hilary Asoko is responsible for the INSET programme and is involved in research into teaching for conceptual development in primary and secondary classrooms. John Leach is working on a project to document primary and secondary pupils' ideas about the nature of science. Phil Scott is co-ordinator of the group.

References

Carey S (1985) *Conceptual change in childhood*, Cambridge, Mass: The MIT Press.

Chalmers A F (1982) *What is this thing called science?* Milton Keynes: Open University Press.

Johnston K (Ed) (1990) Interactive teaching in science: workshops for training courses 9: *Diagnostic teaching in science classrooms*. Hatfield: The Association for Science Education.

Further reading
General
White R T (1988) *Learning science,* Oxford: Blackwell.

Adey P (1989) *Adolescent development and school science,* Falmer Press.

Millar R (1989) *Doing science: Images of science in science education,* Falmer Press.

Leeds National Curriculum Science Support Project. A project jointly supported by Leeds City Council Department of Education and The Children's Learning in Science Research Group, Centre for Studies in Science and Mathematics Education at the University of Leeds (These materials are currently available for Leeds schools and wider publication is proposed).

Conceptual understanding

Driver R, Tiberghien A and Guesne E (Ed's) (1985) *Children's ideas in science,* Milton Keynes: Open University Press.

Osborne R and Freyberg P (Ed's) (1985) *Learning in science: The implications of children's science,* Heinemann.

National Curriculum Council 1992 *Teaching science at Key Stages 3 and 4* (National Curriculum Council, in press).

Procedural Competence

Assessment Matters Number 2 (1990) and Numbers 5, 6, 7 and 8 (1991) School Examinations and Assessment Council.

Teaching science at Key Stages 3 and 4 (1992; in press) National Curriculum Council.

Woolnough B and Allsop T (1985) *Practical work in science,* Cambridge: Cambridge University Press.

Wellington J (Ed)(1989) *Skills and processes in science education: A critical analysis,* London: Routledge.

Developmental influences on learning

Adey P, Shayer M and Yates C (1989) *Thinking science: the curriculum materials of the Cognitive Acceleration through Science (CASE) project,* Kings' College, London. London: Macmillan.

Matching teaching to learning

Scott P H, Asoko H M and Driver R H (1991) *Teaching for conceptual change: A review of strategies,* in Duit R, Goldberg F and Niedderer H (Ed's) (1991), *Research in Physics Learning: Theoretical Issues and Empirical Studies,* Proceedings of an International Workshop, IPN, University of Kiel, Germany.

Needham R (1987) *Teaching strategies for developing understanding in science*, Centre for Studies in Science and Mathematics Education, The University, Leeds LS2 9JT.

The importance of context

Solomon J (1992) *Getting to know about energy in school and society*, The Falmer Press.

Language and Learning

Edwards D and Mercer N (1987) *Common knowledge: The development of understanding in the classroom*, Milton Keynes: Open University Press.

Barnes D, Britton J and Torbe M (1986) *Language, the learner and the school* (3rd edition), Harmondsworth, New York: Penguin.

4 Teaching Science

Phil Ramsden and Bill Harrison

1 Learners and Learning Activities

It seems obvious that learners should have pride of place in any consideration of the best way to teach science. They are the reason for the whole exercise and every teacher knows that no effective learning can take place without their conscious and willing participation. This chapter explores the consequences for teachers of our knowledge about learners and learning.

In the previous chapter we saw that most theories of learning have in common the idea that learning is an interaction between ideas which are already in the learner's mind and new ones introduced by the learning activity. Learning is an active process which starts where the learners are, and any teaching scheme must recognise this:

a) by finding what the learners' knowledge and understanding are and providing learning experiences which start from there;

b) by giving them opportunities actively to test, refine and consolidate their understanding in new learning contexts.

The consequences, for teachers, of such a view of learning are that they must:

i) plan learning activities so that opportunities exist for learners' ideas to be made explicit as a natural part of the lesson structure, so that the process is clearly seen to have the status of 'real learning' i.e. it is an activity which, by its coherence with the rest of the experiences, can clearly be seen as necessary to the learner;

ii) establish a classroom culture in which all contributions, e.g. towards solving a problem, are accepted, and learners feel able to put forward ideas, no matter how strange they may seem, safe in the knowledge that they will not be ridiculed. This is not to say, of course, that ideas should be accepted without question, but rather that it should be made clear that it is the idea which is under scrutiny and not the learner. The role of the teacher is crucial in managing such activities They must judge how far a particular line of discussion can go so that the learners can be challenged sufficiently to think out the consequences of their ideas, without having their confidence so diminished that they no longer participate. As with most human activities there are no simple rules. The extent of a particular learner's confidence in his or her own ideas will vary with the concepts involved and with the context in which they are being used, as well as with all the social factors arising from the working group. Before thinking that to manage such a situation is an impossible task, it is important to remember that skilful teachers with a good knowledge of their pupils are successfully doing it every day;

iii) provide a wide range of learning activities which will encourage learners actively to try out their ideas in new situations in many different ways. The word "investigation" can be used as a blanket term for a group of learning activities which might typically include:
    raising questions;
    planning;
    making observations;
    using practical skills;
    analyzing data and looking for patterns;
    explaining and predicting.

Practical investigations obviously require pupils to be physically active. It is, however, primarily mental activity which is required for active learning. This is not to say that the physical activity is totally independent of the mental: there is evidence that the two may interact to facilitate learning both of concepts and procedural skills.

Practical investigation is by no means the only kind of active learning strategy; other examples would include;

problem solving
small group discussion including poster sessions
drama and role play
surveys and opinion polls
presentations to the rest of the class by individual members and small groups
directed activities relating to text (DARTs)
visits and visiting speakers

Such activities will involve pupils in thinking, creating, predicting, imagining, doing, sharing ideas, discovering, presenting and discussing. A DART, for example, would be a structured activity involving group work, where pupils are asked to engage with a piece of text and process it in such a way that they learn far more effectively from it. (See Davies and Green 1984, Bulman 1985). These activities can only take place effectively as pupils develop their ability to work independently and collaboratively and to take more responsibility for their own learning.

The main aim of all these different activities is the same; to give learners opportunities to develop their understanding by actively engaging with some form of relevant challenging situation.

Learners also need to be able to reflect critically on the success of their efforts and it is in such moments that new learning actually occurs, as the learner rejects one idea which has not 'worked' for another which has. The crucial factor here is the rejection by the learner of the 'old' idea, on the basis of their own experience of using it in a learning activity. Just as the 'old' idea was rejected, or modified, because it did not help to solve a problem or explain a pattern in some data, so the 'new' idea must prove itself in this way, and hopefully be reinforced by being applied to a variety of new situations where it will 'work' better than the 'old' one. Such opportunities for reflection and reinforcement are to be found in the activities listed above, particularly those which involve talking. The National Oracy Project highlighted the importance of talk in developing understanding generally and science is no exception to this.

How then should a science teacher choose which sorts of learning activities to make available to pupils? The choice will depend upon the concepts being taught, the pupils concerned and practical considerations such as time, accommodation and resources. Practical experience tells us that successful teaching occurs when a balanced and varied approach is used in a range of contexts. Later in this chapter more detailed advice is given on preparing and delivering a scheme of work, linked to an example. Before this however it is necessary to look more closely at what is meant by practical work and at the place of practical investigations within it.

2 The Role of Practical Work

A simple analysis of science teaching in the past would identify two main kinds of activity, theory lessons and practical lessons. In some schools the blurring of these artificial boundaries led to a gradual evolution of more effective and interesting science lessons, but in many this was prevented, and the false distinction was reinforced by an acute shortage of laboratories. Theory lessons became an institution, and were timetabled in classrooms where very little practical work could be done. Practical lessons using scarce laboratory time were then so precious that teachers were reluctant to 'waste' them by discussing the theory of what was being done practically.

Even though the majority of science lessons can now be taught in laboratories, the shortage years have left behind a legacy of science teachers who still think in terms of the 'waste of good laboratory time'. In order to confront this understandable but potentially very restricting outlook we need to be clear about what we mean by practical work. The Assessment of Performance Unit (APU) suggested four broad roles for it.

i)   a means of gaining basic laboratory skills e.g. using a thermometer or measuring out a certain volume of liquid.
ii) a means of developing observational skills e.g. observing colour changes or the structure of a cell.
iii) a means of illustrating a particular concept e.g. the movement of a current-carrying conductor in a magnetic field.
iv) a whole scientific investigation.

They carried out a survey of the relative time spent on each role in school science lessons and from their results shown in fig 1 we can see that whole investigations formed only a small part of the practical activities in a science course, with illustrative practical having a dominant role.

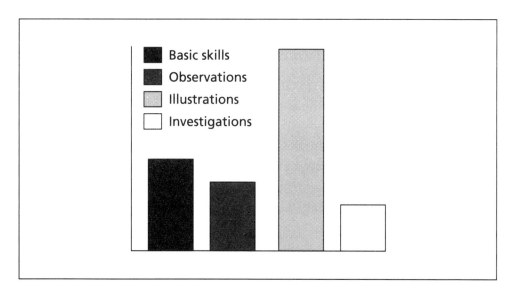

Figure 1 The relative amounts of time spent on different types of practical activity

This survey was done some years ago, before the introduction of the National Curriculum in England and Wales in which Attainment Target 1 is based around investigations. Nevertheless it seems likely that, if the survey were repeated now, it would show only a small increase in the time devoted to whole investigations. Does this matter? Is there a problem?

There are three main answers to these questions.

i)   Whole investigations give pupils their best experience of 'real' science by giving them the opportunity to make use of the science concepts and techniques which they have learnt, so they should form a greater proportion of practical work than they currently do.

ii)   Whole investigations can also fill some of the other roles of practical work such as developing skills. More attention needs to be given to this in planning schemes of work. It is true that they are more difficult to set up and manage than discrete practical exercises, focused on a particular skill. But APU research suggests that the extra effort is worthwhile, because learning will be greater if pupils can see its relevance through a whole investigation.

iii)   The illustrative role for practical work is, however, still a valid one and as these practicals are typically short, they can, if well planned, be very effective. The danger is that if the objectives are not clearly thought out, aspects of practical investigation can creep in, serving only to lengthen the time taken without increasing the effectiveness. The keys to the effective use of practical work and indeed any learning activity are:

- Be clear about what it is intended to achieve
- Choose from the whole range of learning activities the one most likely to be effective: the decision will depend on what is being taught, the pupils' previous experience and motivation levels and on the range of other activities within the same topic.
- Decide if the end is justified by the means: this will often involve teachers in balancing the cost, in time, of setting up a complex demonstration against the benefit to pupils' understanding of actually seeing it 'in the flesh'. Where resources allow, the use of video has a contribution to make, with another trade off to be judged, i.e. the benefit to the pupil of getting a better close up view against the more immediate appeal of a live demonstration.

3 Providing effective learning experiences

Providing effective learning experiences is central to the task of teaching, and consists of several steps.

(i) Planning

This could be the planning of a topic or series of lessons, within an previously planned scheme of work covering a term, a year, or National Curriculum Key Stage. Questions about the aims of the topic, its suitability and the pupils' previous experience should first arise here.

(ii) Preparation

The detailed process of deciding what each lesson will involve, bearing in mind that some allowance for the unplanned will be necessary. Also included here is the production and selection of resources as well as the requisitioning of equipment. In practice this is an extension of the planning process and has only been separated here in order to emphasise the level of detail.

(iii) Managing the learning

This includes teaching the lessons, and organising homework activities: it involves managing pupil behaviour, safety, apparatus and time.

(iv) Evaluation and development

Evaluating and assessing the extent of pupils' learning should happen as much as possible during lessons, though obviously some aspects, such as formal marking, may well occur between lessons. As well as providing feedback to pupils such evaluation may often involve the modification of the original plan to include specific action to be taken with individual pupils or groups. At the end of a topic it should lead to a modified plan being filed for use next time the topic is covered.

Ideally the level of detail in each of the above steps should be just that necessary for effective learning to take place. In practice, this will inevitably be affected by the time available both to plan and to teach the topic. Although there are steps which can only be done by one teacher for a class, there are many others where team work can save time and produce a better range of ideas and techniques, and it is essential that such strategies are exploited fully. In order to discuss the issues involved in each step we are going to look at an example of an actual topic and comment on it step by step.

3.1 A detailed look at a teaching topic

TOPIC: Rusting
YEAR: Y8
SCHOOL: 11-18 mixed comprehensive

The teaching scheme discussed below has been documented as part of a study done by the Children's Learning in Science Project based at Leeds University.

3.1.1 Planning

This is a short section, lasting 3 lessons, within a larger topic called 'Substances,' in a course based around a published lower school scheme. In planning the topic, the teachers identified that few pupils would be familiar with the concept of chemical change and that helping them to develop a good understanding of it represented the central task. They felt that the difficulty of this concept was commonly underestimated in many schemes of work. Rusting was seen to offer a good opportunity to explore pupils' ideas about a familiar chemical change, albeit one about which they would have a wide range of ideas. Through an examination of the factors which are and are not involved in rusting, a good foundation of shared experience could be laid, upon which a better understanding of chemical change could subsequently be built.

Comments

The planning has identified the main aim of the topic (i.e. understanding chemical change) and has then considered

- the difficulty of the key concepts involved.
- the experiences and ideas (particularly ideas which do not accord with the accepted theories of rusting) which pupils will bring to the topic.
- the need to start from some familiar experience of the pupil.

3.1.2 Preparation

The outline plan of the introductory 'Rusting' section covers three lessons.

Three weeks before the first lesson pupils are each to be given a nail and told to take it home, put it in a place where it is most likely to get rusty, and bring it back to their first 'Rusting' lesson.

Lesson 1

Pupils bring back their nails and produce a display including their nails, saying where each has been and what the conditions there were like. This should then lead to a discussion, first in small groups and then as a class, during which pupil ideas of what causes rusting and what rust consists of can be brought out. In addition to water and air being recognised as factors, cold and the presence of other rusty objects are also often cited by pupils, as is the idea that rust is a kind of mould growing from within the nail. As a result of this discussion the teacher has to decide on one of two options.

i)   Groups of pupils can set up experiments to investigate factors from their 'own' nails display which they feel were not clear.

ii) A series of standard experiments can be set up to investigate individual factors which have been agreed by the class as needing further study.

Lesson 2
The outcomes of the controlled experiments are reviewed and some agreement is reached about the essential factors in rusting. The pupils are then asked to look back at the nails display, to discuss some of their ideas about what rust actually is, and to decide how these might be tested.

Lesson 3
Some of the ideas for testing what rust is are tried out and the results discussed, hopefully leading to the idea that it is a new substance formed by chemical change on the surface of the iron. This leads to pupils devising ways in which this knowledge can be applied to prevent rusting.

Comments
In *lesson 1* the teacher must anticipate some of the requirements for the display and of the controlled experiments which might be devised, e.g. more clean nails, salt or salt water, other rusty nails, means of establishing cold conditions and of drying air. The controlled experiments are essential for testing factors of which pupils were unaware during their home experiments: the option to specify particular factors is necessary to ensure that important ideas are not missed. The advantages of relating the next set of investigations to the pupils' own experiments can only be fully exploited if the factors which the teacher wants to be considered are thrown up by the 'nail displays' of pupils within a particular small group. It is worth dealing here with the idea that the second option is actually the fall-back and as such anticipates a failure of the attempt to start from where pupils are. It is important to be realistic in all planning and a fall-back such as this should never be seen as defeatist. In this particular example, each option still seeks to capitalise on the pupils' own experiences and option i) allows for the possibility of doing this in a more elegant and spontaneous way. The fact that this will not always happen must not prevent us planning for it.

Lesson 2 recognises the importance of enabling pupils to move to the accepted view and also illustrates the organisational tactic of taking the pupils' ideas for the next investigations away at the end of one lesson so that the necessary apparatus can be collected together for the next. In lesson 3 pupils' own experience is once again used to lead them to an accepted view and this is then reinforced by asking them to apply it to the very real problem of rust prevention. This is not intended to lead to further investigation but pupils' solutions can be compared to industrial processes used, for example, to protect car bodies, and shown to be in line with the theories of rusting. The reason no further investigations arise from this lesson is simply one of time: judgements have to made on when to bring particular topics to a close to fit them into an overall time plan.

3.1.3 Managing the learning
Some aspects of this have already been discussed. Teachers will also have to be aware of more general aims not usually mentioned in specific plans. The headings below may form a useful checklist.

1. Motivating pupils to work on the topic and maintaining their interest
This is one reason why rusting was chosen and also why each pupil was involved by being given a nail to take home and make rusty. There is a danger that familiar things can be introduced and dealt with in a boring way, so the teacher's skill and enthusiasm in showing the challenges which this everyday situation poses will be crucial.

2. Making sure that all pupils know what is being asked of them
Pupils were each given brief written instructions about the nail: they discussed the tasks in class, both as a whole group and in small groups, before proceeding. While written instructions can be a simple way of achieving this objective it is worth reminding ourselves that their use introduces the factor of the reading ability of the pupils. A useful back up in a situation when verbal instructions have been given or agreed is to ask each pair of pupils to agree between themselves what is required and to check their agreed version with the teacher.

3. Providing differentiated activities or tasks which can be undertaken at different levels of complexity and pursued to different extents
The first task on the conditions for rusting could be addressed at several levels, an example of 'differentiation by outcome'. Some pupils used words like 'damp' and 'cold' while others went on to simulate conditions at the seaside in quite sophisticated ways. The interpretation of controlled experi-

ments in lesson 2 requires a different differentiation strategy which involves the teacher in intervening with individual pupils or groups and leading their thinking by questioning and hints. This is expensive in teacher time, and some pupil-pupil 'tutoring' within groups could be encouraged. Differentiation is one of the key objectives of any teaching scheme and should always be considered, since even setted or streamed groups still contain a mix of abilities.

4. Ensuring that sufficient apparatus and other resources are available

This is achieved in lesson 1 by anticipating what controlled experiments will be needed. In lesson 2, pupils' apparatus requirements for lesson 3 are collected at the end of the lesson. In other cases it might have involved the provision of visual aids for use by the whole class or by particular groups, as part of a differentiation strategy.

5. Encouraging pupils who are becoming disheartened to stick with the task or to modify it if necessary

The teacher, moving around the class and talking to groups and individuals, is keeping a look out for signs of frustration through lack of success. Encouragement, together with hints and practical help, can often be useful, but pupils should also be helped to come to terms with the idea that investigations can fail, but that learning from our mistakes can be viewed positively. The relationship which a teacher has built up with pupils is particularly important here.

6. Dealing with dangerous or silly behaviour

The opportunity to handle apparatus and to move about increases the scope for bad behaviour; science teachers need constantly to be on the lookout for this. Interesting and well planned lessons, where spare apparatus is carefully managed, will lessen the risk but not remove it. Most bad behaviour in laboratories is unsafe and can be dealt with as a breach of safety rules. The first priority is to stop the behaviour and then to confront pupils with the possible consequences of their actions. It is essential that clear safety rules have been agreed and noted by the whole class and it helps if among these is the need to stop and listen when suddenly asked to do so by the teacher in an emergency. Pupils should be encouraged to see safety training as a necessary and natural part of their work which enables them to progress to do ever more complex and hopefully interesting practical work without unacceptable risk. One aspect of involving pupils in the planning of investigations is that they should be encouraged to identify potential hazards and agree on precautions to be taken, thus making it easier for the teacher to challenge any subsequent dangerous behaviour on the ground that it is a breach of reasonable rules, jointly agreed.

7. Gathering evidence of how well pupils are understanding the work
This is another spin-off from talking to pupils and groups and it is part of the data which allows teachers to make adjustments to the task as it goes along. Marking of pupils' written work will provide a second chance to form judgements about their understanding. Marking should normally be done within an overall departmental and school policy and should be against agreed criteria which may be specific for a particular task or more general for others. In each case, pupils should be aware of the criteria being used and hence the meaning of any grade given. Pupils should also be clear about the teacher's expectation of the action they should take as a result of the marking, and that this will be followed up by the teacher. This need not be as complex as it may sound. On many occasions it could involve the teacher checking records and seeing some pupils to discuss a particular issue. The important thing is that pupils should be given the consistent message that marking is not the end of an episode of learning but one stage in it, and that there is value in them following the suggestions contained in the teacher's comments.

8. Encouraging pupils to think critically about the observations they are making
Once again sensitive questioning as the teacher moves around is the key, but some questions can be included in the written material. Allowing time for pupils to report their results to each other offers another route for critical analysis, but teachers need to manage such situations to avoid undue stress being put upon individuals.

9. Ensuring that pupils make sufficient records of their findings and ideas
Decisions need to be made about the purpose of the records and pupils should be a part of this process so that they see the need for the recording they are doing. This will vary in different topics. In the rusting topic, pupils should see why some recording is necessary to make the rusty nail display intelligible to others. Records of their theories of rusting are necessary, for the teacher to look at and for the pupils' own future reference. Once again departments should have overall policies about recording, and the form of pupils' reports on their investigations. (See page 267) The overriding criterion should be fitness for purpose, rather than preparation for future study at a higher level.

10. Providing opportunities for pupils and groups to share their ideas and findings with the whole class

As well as leading to more critical analysis, this forces pupils to think about what they have done and to examine the logic of what they are going to say or write. The importance of talk as a means of gaining and deepening understanding has been demonstrated by studies such as those done by the National Oracy Project. In science we need to make more use of it in discussion and reporting back. Careful management may be needed to avoid this being too much of an ordeal for some, but it is a vital skill to develop. The work of the ATLAS (Active Teaching and Learning Approaches in Science) Project has many useful suggestions to make about this and most other forms of active learning (Harrison; in press).

11. Encouraging pupils to monitor, evaluate and assess their own learning
This will link into the teacher's own evaluation; in addition to it being a check on their own perceptions and judgement it is an important skill which pupils should develop, since it is essential for advanced study as well as in everyday life and employment.

3.1.4 Evaluation and development
A great deal of evaluation of learning went on during the topic – necessarily so, as it was used to modify the lessons. For example, teachers had other options for lesson 1, one of which was to use the rusty nails display to arrive at the accepted view of what causes rusting. This was not followed, as the pupils had not covered the range of conditions needed in their home experiments and had not separated variables enough for judgements to be made. The teachers therefore decided to ask pupils to set up controlled experiments to sort out and investigate variables. Such evaluations are a vital part of teaching a topic as are those of the development of pupil understanding, which may lead to extra emphasis being placed on a particular aspect or to another being dealt with more quickly. However creative and flexible a topic plan, there will always be a limit to the modifications which are possible, and this is why a more general evaluation at the end of the topic is needed. This has two main functions: firstly to decide what further learning experiences on the concepts involved are needed by the pupils; and secondly to decide how the particular topic could be dealt with next time around. In this example, the rusting topic leads into a more general one on chemical change, so a crucial evaluation is the extent to which pupils have understood this concept and how much

further exemplification is needed. Other less fundamental but still important evaluations are whether pupils responded well to the responsibility they were given for planning their own investigations and whether they were always clear about what was expected of them. The topic will be covered 'next time' with different pupils, but some planning decisions can nevertheless be made. For example, if even with a well motivated group the practical problems of taking the nail home, keeping it safe and returning it on time were to prove too great, another strategy would be needed.

4 The Science Teacher

Famous scientists have publicly paid tribute to individual science teachers who motivated and inspired them to study science. What qualities did these inspiring teachers of the past possess? It would be all too easy to regard such teachers and their teaching as a feature of a bygone age of selective schools and formal teaching methods, and to dismiss the 'teacher factor' as being largely irrelevant to today's science education scene, with its national curricula, whole school approaches and departmental teams. To do this would be wrong because the skill, personality and integrity of individual science teachers and the relationship which they foster with their pupils is still at the heart of successful teaching and learning.

Television and the media will have given many of today's pupils a wider awareness of scientific topics such as astronomy, medicine, energy, force and materials, and issues such as pollution, global warming and the population explosion. They will therefore come to science lessons with a considerable background of experience and language which can then be more clearly related to science concepts. Despite this increased experience, their science teacher is still likely to be the first and only representative of the scientific community whom they come to know in any depth. The fact that teachers might not regard themselves as 'scientists' because they are not involved in scientific research is a subtlety lost on most pupils and so, whether they like it or not, science teachers are scientists to their pupils. For this reason, if no other, it is vital that science teachers show that they have genuine interests in scientific issues and topics. This should be linked to efforts to give pupils as many opportunities as possible to develop their own interest in science 'for its own sake'.

The applications of science, suitably woven into science topics, can of themselves make a good case for the value of science as servant of society, and this case is now well made with most pupils. What is much less secure in pupils' minds is any appreciation of science as an intellectual activity which can be pursued for its own sake, simply as a means of satisfaction and enjoyment. Such an appreciation will, for many pupils, depend on their teacher's own attitudes to science and the interest and enthusiasm demonstrated to pupils. One difficult area here is that of specialisation. Pupils will all too quickly appreciate the practical need for specialisation as they themselves are introduced to the vastness of science. It is vital however, that science teachers encourage them to have as wide an experience of science as possible, as part of their general education and as a better preparation for possible future scientific specialisation. The importance of scientific discoveries being made in areas not neatly covered by current specialism should always be emphasised.

Teachers will often themselves have specialised at advanced or degree level, but it would be undesirable for pupils to emulate this specialism at too early an age. This is less likely if science teachers can show themselves to their pupils as scientists who have a general interest in all science.

Given that teachers are seen as scientists by their pupils, it follows that they will also have a crucial role in showing pupils what science is. Among the characteristics that the famous scientists, mentioned in the introduction, found most attractive in their science teachers were enthusiasm for the subject and an insatiable curiosity, which we might have predicted, and the ability to say "I don't know the answer", which we might not. Yet this last factor was seen as crucial in establishing the integrity of science for these scientists as young people. Honesty with regard to scientific results is also a crucial way of showing pupils the value which is put on the process of investigation as a genuine creative activity. All these things build up a picture of science in pupils minds as a worthwhile activity which can offer fulfilment as well as challenge. Having a fully developed sense of what science is will obviously help teachers to encourage some pupils to pursue it further but it is also at least as important that those who do not do so leave school with clear ideas about science, what it is, why it is studied and what its limits are. There are many reasons for this, but probably the two most important are that they should not fear science out of ignorance and that, as voters in a democracy, they should be able to make enlightened decisions about the funding and application of it. Much of the responsibility for these lofty ideals must be carried by individual science teachers and cannot be devolved onto learning resources or apparatus, however ingenious and exciting.

Phil Ramsden is an Advisory Teacher for Science in Sheffield, having taught science for 17 years. He has been involved with various science teaching projects including Nuffield, SCISP and balanced science in TVEI.

Bill Harrison is head of the Centre for Science Education at Sheffield Polytechnic, working in initial training and INSET. He was a member of the SATIS central team and director of the ATLAS (Active Teaching and Learning Approaches in Science) and PSI (Problem Solving with Industry) projects.

References and Useful Reading

Bentley D and Watts M (1989) *Learning and teaching in school science*, Open University Press.

Part 1

Davies F and Green T (1984) *Reading for learning in the sciences*, Oliver and Boyd.

Bulman L (1985) *Teaching language and study skills in secondary science*, Heinemann Educational.

McConlough E (1991) *Group forensic work in a murder enquiry,* SSR vol 72, no 260, p 142.

McMurdo A (1990) *Taking the plunge with role play. Departmental in service using SATIS materials*, SSR, Vol 71, no 257, p 131.

Campbell R (1988) *PGCE students inject drama into science teaching*, SSR, vol 70, no 250, p 114.

Part 2

Saunders A (1990) *Do practical subjects encourage understanding of science?* SSR, Vol 71, no 257, p 125.

Hodson D (1990) *A critical look at practical work in school science*, SSR, Vol 71, No 256, p 33.

Hodson D (1992) *Redefining and reorienting practical work in school science,* SSR, Vol 73, No 264, p 65.

Part 3

Harrison W (In press) *Active teaching and learning approaches in science*, Collins Educational.

Sands M K and Hull R (1985) *Teaching science – A teaching skills workbook*, Macmillan Educational.

Secondary Science Curriculum Review (SSCR) (1985) *Better Science: Learning how to teach it*

SSCR (1985) *Better Science: Approaches to teaching and learning.* Both ASE/Heinemann Education for the Secondary Science Curriculum Review.

Domenech A & Casasus E (1991) *Galactic Structure: a constructivist approach to teaching astronomy*, SSR, Vol 72 No 260, p 87.

Lock R (1990) *Open ended problem solving investigations: what do we mean and how can we use them?* SSR, Vol 71 No 256, p 63.

Bland M and Harris G (1989) *Peer tutoring*, SSR, Vol 71 No 255, p 142.

Foster S (1989) *Streetwise physics*, SSR, Vol 71, No 254, p 15.

Haines C (1989) *Experience with ILPAC: individual learning via supported self study*, SSR, Vol 70, no 253, p.103.

Byrne M & Johnstone A (1988) *How to make science relevant*, SSR, Vol 70, No 251, p 43.

Arnold M & Miller A (1988) *Teaching about electric circuits: a constructivist approach*, SSR, Vol 70, No 257, p 149.

Smith N (1988) *In support of an applications first chemistry course: some reflections on the Salter's GCSE scheme*, SSR, Sept 1988, Vol 70, No 250, p 108.

Part 4

Bodmer Sir W (1989) *Scientific literacy for health and prosperity*, SSR, Vol 70, No 253, p 9.

5 Assessing and Evaluating in Science Education

John Skevington

This chapter aims to provide an outline of the organisation of assessment in the secondary phase of education in England and Wales, particularly as it applies to science, and to highlight some of the important issues (with some practical ideas) which should guide assessment if it is to make a worthwhile contribution to teaching and learning.

1 Introduction: Why Assess?

We have been used in secondary schools to externally administered terminal examining, to provide a qualification for entry into the next stage of education or the world of work and, as a by-product, an informal indication of the effectiveness of teaching. Additionally, school-based examinations have provided a basis for decisions on pupil grouping and option choices, and class tests have been used as a means of diagnosing pupil progress on a more frequent basis.

With the introduction of the National Curriculum and its associated testing, there is a danger that a narrowly focused, externally imposed assessment system will produce an undesirable back-wash effect on school-based assessment. If this is to be avoided, teachers in science departments need to decide exactly what they believe are the outcomes of science education – which may well go well beyond the transmission of a body of knowledge and a particular set of skills – and develop the assessment system needed to reach these goals.

The ASE Policy Statement (ASE 1992) suggests three major purposes of assessment:

to help learners and teachers;
to provide relevant information to others;
to help institutions to perform better.

These purposes are not at odds with the statutory purposes of assessment but they should go well beyond them. Rowntree (1988) provides a useful discussion of the 'why' of assessment. In-service materials such as those produced by the Industrial Society (1989) and the Centre for Assessment Studies at the University of Bristol (1992) can provide a framework for developing a whole school policy.

Assessment which helps teachers and learners is generally known as *formative assessment*. Its outcomes are used diagnostically. The areas of achievement and weakness of pupils are determined, so that decisions can be made about the next stages of learning and remedial action can be taken where necessary.

Assessment which is used to inform others through some form of reporting may be formative or summative. A *summative assessment* is made at the end of a learning unit or course. It provides information on the level of achievement reached by learners. The report may be teacher generated, a statement of level derived from an externally imposed system such as National Curriculum tests, or in the form of a qualification such as GCSE. However, the results of summative assessments are frequently used to assist in making choices of courses to be followed, and can thus be considered as having a formative purpose.

As a general principle, summative assessments should only involve knowledge or abilities which will be relevant in the future. For example, there is little point in assessing a particular aspect of a course which is used to develop understanding but has no application beyond this.

The use of the results of assessment in planning and reviewing teaching and learning, to improve the performance of the institution and its members, is what we shall term evaluation. This is the usual usage of the term in this country. The terms assessment and evaluation tend to be interchangeable in American publications.

2 An Overview of National Assessment Systems

2.1 11–16

The 1988 Education Reform Act, and the various Orders on reporting and assessments which have sprung from it, lay down minimum requirements for schools. These are outlined in Table 1. (DES, 1992)

Which parts of the report will be derived from whole school information collection?

Which of the reported information will be derived from teacher assessment in science?

Table 1 Information which must be reported to parents

REQUIREMENT	*All* pupils	All of compulsory school age	All at end of key stage	All at end of key stage 1	All at end of key stage 4	All over compulsory school age	All entered for A/AS examinations
General progress △	O	O	O	O	O	O	O
Arrangements for discussion △	O	O	O	O	O	O	O
Particulars of all subjects and activities	O	O	O	O	O	O	O
Public examination results (if applicable)	O	O	O	O	O	O	O
Comments on *all* NC subject studies △		O	O	O	O		
Attendance record △		O	O	O	O		
NC assessment by subject and PC level ø			O	O	O		
Comparative information about achievements of pupils of the same age in the school △			O	O	O		O
Score in Ma3 (arithmetic) △				O			
Reading accuracy score △				O*			
Result of spelling test △				O+			

△ New requirements
ø Only when statutorily assessed
* Only pupils with reading abilities around the average (level 2)
+ Only pupils with spelling abilities above the average (levels 3 or 4)

82

Any assessments which teachers make on top of the minimum necessary to meet these requirements must form part of a coherent and systematic programme, if the additional effort involved is to be worthwhile. We will consider what this involves later.

2.1.1 Key Stage 3

Assessment at the end of Key Stage 3 is by timed written tests, which provide the levels in Attainment Targets 2 to 4. The tests are externally set, teacher marked and externally moderated, by a 'Quality Audit' process which is managed by the examination boards. The level for Sc1 is arrived at by teacher assessment and it is also externally moderated as part of the quality audit.

The levels achieved in each attainment target are averaged to provide an overall subject level. Teachers are required to provide an assessment of the levels for attainment targets 2 to 4 but, apart from pupils whose absence from the tests is authorised, no statutory use is made of these results.

2.1.2 Key Stage 4

Assessment at age 16 (the end of Key Stage 4) is primarily by means of GCSE. Most syllabuses provide a framework for the assessment of Sc1 which comprises the assessed course-work component of the scheme. Some are organised in a modular fashion and provide for an element of assessment during the course which, in addition to contributing to a pupil's overall summative assessment, can also be used for formative purposes.

(There has been a marked increase in the development and use of modular schemes of teaching and assessment since the early seventies, and particularly since the introduction of GCSE. For example, NEA Science (Modular) grew from 30 000 to 120 000 from 1988 to 1992. The use of end-of-module or periodic testing was felt to provide a motivation for pupils, especially the less academically inclined. The nature of such courses also matched a shift away from a content based idea of science education towards the view that the processes and skills of science should be given an increased emphasis, because periodic assessment tended to reduce the emphasis on long term memory. This move was accompanied by an increase in the amount of teacher assessment which contributed to the final grade, both practical skills and abilities which could not be appropriately assessed by a time-limited terminal examination, and, in some schemes, the assessment of knowledge and understanding.

The introduction of the National Curriculum and recent policy changes, which have increased the weighting of terminal assessment in GCSE, have tended to reduce some of the advantages which were claimed for modular schemes. The specification of a relatively large body of knowledge and understanding in the NC appears to have eliminated the options for choice in science courses. The contribution of teacher assessment to the final reported level has been limited to the assessment of Sc1, and the end of module tests are allowed to contribute no more than 25% to the final level.)

The teacher assessment of Sc2, 3 and 4 for formative purposes, and to provide the basis of the reports which parents are entitled to request at any time, is left to the individual school to organise. It would seem sensible to try to develop a consistent assessment programme for Key Stages 3 and 4. There are two schemes which provide 'off the shelf' assessment programmes which go some way towards meeting this need.

The *Graded Assessment in Science Project* (GASP) (ULEAC 1992) provides a five year programme of assessment and recording of achievement which is supported by published materials. The final two years of the scheme lead to a GCSE administered by ULEAC. *The Suffolk scheme* (MEG 1992) is a GCSE syllabus administered by MEG but it includes an optional Year 9 programme which can provide the basis for teacher assessment in the final year of Key Stage 3.

Both schemes provide detailed criteria for teacher assessment of content, derived from the Programmes of Study and the Statements of Attainment, and pupils are assessed on a 'can do' basis against them. Even if a department decides to use a different GCSE course, the criteria developed by these courses can form the basis of an assessment scheme.

The GCSE will only provide certification for pupils who achieve level 4 and above. For pupils who are not likely to achieve level 4, the examining boards provide alternative schemes which involve varying amounts of teacher assessment.

2.2 Assessment Post-16

The picture post-16 is much more complex than that created by the National Curriculum and, at the time of writing (July 1992) is not a matter of law. It is however subject to regulation and control by SEAC and the NCVQ, the two organisations which administer government policy. The developments at this level are described in detail in chapter 16.

In 1996, A levels are to become subject to a set of general principles which have been developed by SEAC and accepted by the Secretary of State.

It is likely that a set of subject cores will be developed for the main subjects, including the sciences, and these will specify objectives and common areas of content, which must be included in all syllabuses for that subject.

The alternative 'vocational' path has traditionally been the preserve of further education colleges, although such courses are increasingly being offered in schools. There is a range of vocational courses administered by City and Guilds and the Business and Technician Education Council (BTEC). The courses are presently being rationalised by the NCVQ. (See page 295)

3 Principles of Effective Assessment

Although the area of external testing and qualifications remains one of constant change and uncertainty, there are recognisable principles upon which assessment should be founded. It should:

- assess the outcomes of teaching and its associated learning;
- reflect and support the aims and objectives of a course;
- support the teaching and learning strategies employed without distorting them;
- provide sufficient evidence to enable a professional judgement of the learner's achievement to be made;
- form an integral part of the scheme of work, (so should therefore be developed alongside it rather than as an 'add on').

3.1 Starting from aims and objectives

The starting point is the aims and objectives of the science course. We need to have a firm understanding of what we want pupils to achieve so that we can design methods of testing how successful they have been. Some aims and objectives may be supplied by an external syllabus, but the school ethos and the outlook of the teachers in the science department may result in further objectives being added.

The National Curriculum does not provide stated aims. The accompanying Orders state that the statements of attainment

"*provide the objectives for what is to be learned in each subject*".

When developing schemes of work it is useful to have a more general view of the objectives.

Table 2 Objectives from the National Criteria for GCSE Science

Scientific Investigation

3.3 The assessment objectives in paragraph 3.4 – which derive from the science Order – allow candidates to be given a level on Attainment Target 1 for all schemes designed to assess National Curriculum science. The objectives equally apply to non-National Curriculum syllabuses.

3.4 Candidates should explore and investigate the world of science and develop a fuller understanding of scientific phenomena, the nature of theories and the procedures of scientific investigation. Their assessed activities should require a progressively more systematic and quantified approach which develops and draws upon an increasing knowledge and understanding of science. Candidates should be required to use the knowledge, skills and understanding specified by the syllabus to plan and carry out investigations in which they:
   (i)   ask questions and hypothesise;
   (ii) observe, measure and manipulate variables;
   (iii) interpret their results and evaluate scientific evidence.

Knowledge and Understanding of Science

3.5 Syllabuses designed to assess National Curriculum science should include the knowledge, skills and understanding specified by Attainments Targets 2, 3 and 4 and the associated Programmes of Study.

3.6 For other science subjects, the required knowledge, skills and understanding must be specified by the syllabus.

3.7 Within the knowledge, skills and understanding specified by a syllabus, candidates should demonstrate the ability to:
   (i)   communicate scientific observations, ideas and arguments effectively;
   (ii) select and use reference materials and translate data from one form to another;
   (iii) interpret, evaluate and make informed judgements from relevant facts, observations and phenomena;
   (iv) use science to solve qualitative and quantitative problems.

Look at the objectives from the National Criteria for GCSE Science (SEACb, 1992) as shown in Table 2.

They are considerably briefer than those provided by the original GCSE criteria for science (published in 1987) and used by science syllabuses prior to the National Curriculum.

What additional objectives would you consider important for science courses in Key Stages 3 and 4?

The overall objectives of a course are often too broad for assessment purposes. They need to be related to specific parts of the content of the course or to particular skills or processes. This involves establishing suitable criteria for assessment, often called behaviourial objectives, which specify what learners are expected to do in order to demonstrate achievement of the objective. The discussion which follows uses the context of the National Curriculum to develop these ideas, but they are equally applicable to any course.

Sometimes the statements of attainment in the National Curriculum are difficult to use as criteria for assessment because they are so broad. For example, Sc2 6a

"Pupils should be able to relate structure and function in plant and animal cells"

could not be sensibly assessed by one exercise. The problem here is that there is a very wide range of cell types in plants and animals, each with a different structure and function. If pupils are to be asked:

Describe how the structure of eg. a muscle cell and/or a nerve cell and/or a mesophyll cell is related to its function....

then they – and their teachers – clearly need to know, in advance, which types of cell are included on the assessment agenda. It is, however, possible to produce assessment items for which the above SoA is an adequate criterion, eg.

*Describe how the structure of **one** type of animal cell and **one** type of plant cell enables the cell to do its job effectively.*

Alternatively, the assessment item could present information about the structure and function of particular cells – preferably cells which pupils are unlikely to have met during their course – and ask pupils to explain how their structure enables them to do their jobs.

For the purposes of formative assessment, a more sensible approach would, of course, be to collect evidence over a range of activities and over a period of time. To provide adequate opportunity for such assessment, it is important to identify the points in the scheme of work where the content implied by this statement is addressed. The appropriate assessment techniques can be related to the learning activities, and the contribution of the assessment of this particular area of the curriculum to the whole can be built up.

Other SoAs present problems as assessment criteria because they do not specify what pupils need to do in order to demonstrate achievement, a disadvantage for teacher and pupil alike. An example of this is Sc4 3b.

> *"Pupils should know that there is a range of fuels used in the home."*

A pupil writing or saying "There is a range of fuels used in the home" actually meets the requirements of the SoA but is scarcely convincing. The statement needs rewriting before it can be used as a criterion for assessment.

One way of tackling this problem is in the *Common Themes* developed by NEAB/WJEC/ULEAC for their GCSE Science Frameworks. They comprise a comprehensive set of assessments including not only necessary interpretations of National Curriculum SoAs but also a comparable set of statements for concept areas included in the PoS for which there are no SoAs. For example the section on fuels includes the following:

> *at level 3*   *To make things hotter we can:*
> - *burn gas, coal or other fuels*
> - *use electricity*
>
> *at level 4*   *Coal, oil, gas and wood are all fuels. They release energy when they are burned.*

The required knowledge and understanding , though still in the form of substantive statements, is now non-trivially testable as recalled knowledge, eg:

> *Write down the names of **two** fuels people use in their homes.*

> *One way of heating water at home is to burn a fuel such as gas. Write down another way of heating water at home.*

> *Gas, wood and oil are all **fuels**. Explain what this means.*

Specifying the syllabus, wherever possible, in the form of statements, particularly ones expressed in the language which pupils are expected to understand and use, makes it clear to both teacher and pupil what is required.

3.2 Using an assessment instrument which is appropriate for the purpose.

3.2.1 Knowledge and understanding

There is a wide range of methods of assessment by which knowledge and understanding can be assessed. The reader is recommended to consult a text such as Fairbrother (1988) for a complete discussion. The principle of 'fitness for purpose' should guide the selection of method. Decide on the purpose of the assessment and the nature of the objective which is being assessed, and then select a method which will enable you to assess that objective in as valid and reliable a manner as possible. By valid, we mean that the assessment measures what it is intended to measure, as defined by the assessment objectives; by reliable, we mean the consistency with which the assessment measures achievement.

Validity is, in part, achieved by carefully matching the design of the assessment item to the assessment objective. Using a range of appropriate assessment methods will also tend to increase it. However, this needs to be balanced against the increase in time demanded and a possible reduction in coverage of the objectives.

Reliability is increased by carefully defining the desired outcomes of the assessment; in other words having a clear idea what aspects of pupil performance will provide the information that is needed.

(In practice there is, unfortunately, sometimes a conflict between the demands of reliability and validity. It is this conflict which underlies the disagreement about the relative contributions which course-work and terminal examinations should make to pupils' assessment at GCSE. Course-work assessment is usually based on a large sample of assignments of different types and often of a more realistic nature than is possible in examination papers, so is likely to be more valid in terms of overall aims and objectives. Ensuring that different groups of pupils undertake comparable assignments under comparable conditions, and that the criteria for their assessment are consistently applied is, however, far more difficult with course-work than with final examinations. The introduction of GCSE saw a move towards a greater emphasis on course-work assessment. The National Curriculum attaches a greater importance to reliability than to validity and has reversed this trend.)

The assessment should be capable of being operated within the time and the resources available. This is especially important if assessment is not to make disproportionate demands and so distort teaching and learning.

The next stage is to identify exactly what aspects of pupil behaviour will allow a judgement to be made. If the objective requires recall of knowledge then the activity must involve the communication of the recalled knowledge in some form, whether oral, written, pictorial or by demonstration in some practical activity.

Many statements of attainment require that pupils understand a concept or relationship and we need them to demonstrate this understanding in some way at some level. In part, this level is set by the level and range of the concepts involved. Within the National Curriculum it is possible to identify a progression in the complexity of understanding required of pupils. This can be because of the links to other concepts which are implicit in the understanding demanded. For example, Sc4 7d requires an understanding of 'the quantitative relationships between force, distance, work, power and time'. This goes beyond the ability to substitute the correct quantities in the appropriate formula and carry out a calculation correctly. It embodies concepts of energy transfer, the idea that energy can be quantified, and an understanding of some of the effects of forces.

The problem is further compounded by the development of a more quantitative approach as one progresses through certain strands. We should, however, take care not to confuse the ability to use mathematics, important though this is in many aspects of science, with a comprehension of the relevant scientific concepts. For this reason, many conventional mark schemes allow for credit for understanding and application of scientific ideas, even if the pupil makes numerical errors.

It is especially difficult to assess understanding in timed written tests. One word or short answers are poor indicators, as are methods where rote application of techniques is possible. One way in which it can be assessed is via items which present information about situations, phenomena or devices which are not specified by the syllabus, and which are unlikely to have been used in the teaching or learning of the course. Pupils are then asked to demonstrate their understanding of science concepts which are specified by the syllabus by using them to explain, evaluate, or in some other way make sense of the presented information. Of course, the assessment of understanding lends itself to teacher assessment. A range of activities can provide evidence for understanding. Investigational work, hypothesising and the design of investigations are suitable if properly set in the context of the knowledge and understanding required by the other attainment targets.

Spider diagrams or concept mapping allow pupils to demonstrate their understanding of links between different areas of the subject. Pupils can be asked to identify the key concepts in their diagram or for the reasoning behind the links which they have made. An example of such a spider diagram is given in figure 1.

Asking pupils to write questions is another technique for testing understanding. There are various ways of ensuring that the questions do not test simple recall. We can ask for questions which begin in a specific way such as 'what if' or 'how would'. A stimulus such as a diagram of a physical situation or a numerical answer with a specific context can be an effective way of starting. Pupils can be asked to produce suitable answers to their own questions as an extension to the activity.

These and other techniques are discussed in much greater detail by White and Gunstone (1992)

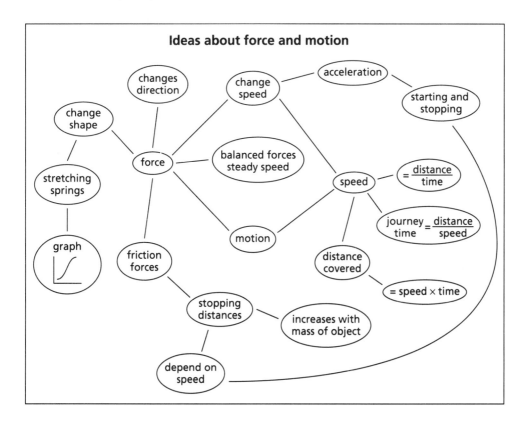

Figure 1 A spider diagram of force and motion

To complete this consideration of how effectively to assess knowledge and understanding, a number of other issues need to be considered.

Firstly, whereas formative assessment might legitimately focus attention on the specific details of the particular learning experience which pupils have encountered, this is inappropriate for summative assessment. Summative assessments should be concerned with the science concepts themselves, not with the ways in which pupils acquired that understanding. Assessing such understanding may, as indicated above, involve asking pupils to apply science concepts to specific situations but the details of these situations should be supplied in the assessment items themselves; they should not be part of what pupils are expected to recall. This principle, consistently applied, can make an enormous contribution to reducing what might otherwise be a hugely information-overloaded National Curriculum.

Secondly, the National Curriculum not withstanding, the degree of difficulty of an assessment item depends not only on the level of the concepts involved, but also on the language used – particulary the non-scientific 'carrier' language – and on the context within which pupils are asked to demonstrate their understanding. A short paper produced by the British Association of the Teachers of the Deaf offers very useful advice on the 'Language of Examinations'. Even very experienced examiners find it difficult to predict the effect of a particular context on the difficulty of an assessment item. For this reason, examining groups intend, within the limited range of levels targeted by each differentiated paper, to use marks rather than a strict criterion-referencing of items to levels. This approach will mitigate not only any unanticipated effects of context on the difficulty of items but also any inappropriate attribution of levels within the National Curriculum itself (levels which are hardly scientifically established!).

Thirdly, the requirement for differentiated papers, targeted at a relatively narrow range of levels, presents teachers with real problems about what the teaching/learning agenda for particular pupils should be. Though pupils are normally expected to follow the PoS at each key stage, the National Curriculum itself recognises that not all of the PoS for a key stage is appropriate for all pupils, and that aspects of earlier key stages may also need to be included. Furthermore, whilst it is clearly undesirable to deny pupils access to concept areas from which they might derive educational benefit, it is equally undesirable that the examinations for which they enter should provide them with no opportunity to demonstrate what they understand of concepts specified as being at a higher level than those targeted by the paper.

Science teachers will not be able to avoid having to make some very difficult decisions.

3.2.2 *Investigations and practical work*

The introduction of GCSE demanded that teacher assessment of practical work be employed as the most appropriate means of measuring achievement of laboratory and experimental skills. Syllabuses provided lists of criteria against which pupil's work could be measured, either within the context of a whole experiment or as an isolated exercise. Such skill-based assessments can be an appropriate formative tool. However, Sc1 has a particular view of the nature of the experimental work which demands that teacher assessment for the purposes of the National Curriculum is in the context of whole investigations. An investigation requires pupils to:

- *formulate questions based upon existing knowledge and understanding, which they can test:*
- *put together a sequence of investigative skills (such as identification of key variables, observing and measuring) into an overall method to solve a problem;*
- *evaluate their findings in the light of the original problem;*
- *define and develop the way they tackle problems, and what they know and understand. (SEAC, 1992a)*

For the purposes of assessment, the scientific knowledge and understanding which provide the context for the investigation should be of a comparable level to the investigation. Thus it is unacceptable (and probably impossible) for a level 6 or 7 investigation to be based upon an idea which is expressed in a level 2 Statement of Attainment.

In the context of teaching and learning, we need to expand ideas of the purposes of practical work to embrace what to many teachers is a novel area. Furthermore, we need to balance the demands of this new area against the need to develop more conventional practical skills, as well as knowledge and understanding, in the context of practical situations.

> Find or invent a practical activity which could serve as a vehicle for an investigation at a specified level e g. 6, 8, 10.
>
> What sort of knowledge and understanding is involved and at what level? What sorts of question or hypotheses might pupils pose?
>
> What variables are or might be involved and what is the nature of these variables?
>
> Could the activity provide an opportunity for generating quantitative predictions? If so what predictions might pupils make?

Analyzing possible practical activities in this way will enable teachers to identify suitable vehicles for investigations as well as the range of levels which might be achieved.

3.3 Managing effective assessment.

Knowing why and what we need to assess is one thing. Being able to carry it out, record the results and maintain a record of the evidence used for making judgements for later use is the key to making it work.

The classroom or laboratory needs to be organised in such a manner as to facilitate assessment. This means that the teacher needs to be able to focus attention on one group of pupils whilst assessing them, yet still be able to see what else is going on, and respond to other members of the class if necessary. Pupils need to have a clear idea of the structure and pattern of their work, so that they can direct their own learning for a suitable period. To facilitate this, resource materials should be organised so that pupils are able to carry on with their work with a minimum of attention.

For some groups such self-study may not be possible. Resources may be inadequate and the nature or abilities of the pupils may not allow it. In such circumstances assessment activities will have to be restricted or adapted. The teacher has to make a decision as to what is in the best interest for the development and learning of the pupils.

Assessment is liable to be more successful if pupils understand what is expected of them, both in terms of what they have to know or be able to do, and of the performance that is needed to demonstrate their achievement. This will involve writing the assessment objectives in terms which pupils can understand. For formative assessment, this will enable pupils to participate in the process, which in itself can help relieve the burdens on the teacher. This approach can also be used to facilitate the production of a record of pupil achievement, which is a further strand in the policy of most schools.

The need to assess to externally set standards and the legal responsibilities involved in summative assessments will mean that self-assessment by pupils is unlikely to be an acceptable basis for summative assessment. At the relevant stages there is a requirement for schools to conduct a process of internal moderation and standardisation to ensure that all teachers are assessing to a common standard. This process is made much easier if the major part of this standardisation can be achieved at the stage of planning assessments. Discussions between teachers when designing assessments and their associated behaviourial objectives should help ensure that assessments across the department are to a uniform standard.

3.4 Sensible record keeping

There is a risk that the conscientious and enthusiastic teacher, aided by like minded pupils, will generate so much information during the process of assessment that much of it will be useless, because nobody will be in a position to make use of it! The key to sensible record keeping is to decide at the outset just what information will serve the purposes which the school and department policies have laid down and record only that information. Much assessment will be informal, and feedback to pupils will be verbal or by written comment against their work. In many cases it will not be necessary for the teacher to keep a separate record of this. Assess when there is a reason to assess and record when there is a reason to record.

What records will be necessary to enable you to meet the regulations on reporting to parents and to make the necessary assessments for the end of the key stage? Remember that you may need to produce the evidence which you used to make judgements about particular pupils.

What records will you need to make to keep track of pupil progress? Where would be the best place to keep such records? What records will you need to keep to enable you to send a useful report with the pupil on transfer to another group or to another school? (Remember that you may be required to produce such a report at any time during the year.)

Sensible record keeping also requires that as far as possible the recording is done at the same time as the assessment. This reduces the time involved and makes for more reliable recording in situations which may not produce a permanent record of the pupil's work. The record must mean something in terms of the objectives of the assessment. It is unlikely that the conventional 'mark out of ten' will provide adequate information for arriving at judgements of levels.

Some activities will allow assessment of the whole class to take place during the same lesson. For others a proportion of the pupils will be assessed. It is more effective to maintain an on-going record on a whole class sheet rather than on individual pupil records. An example of such a sheet is given in figure 2. The date is there to trace pupil's work if it is needed as evidence; the criteria provide a reminder of the basis for the judgements made by the teacher and the comments can supply further evidence for making judgements.

TOPIC	Forces and Motion Sc4	DATE	22/1/92
WKSHT/EXC CODE OR TITLE	Spinner Investigation Sc1 Worksheet 9.49	TEACHER	J. Williams

CRITERIA			LEVEL
CRITERIA	1	Suitable question/hypothesis posed	2–3
	2	Based on experience P/scientific knowledge S	3 → 5
	3	Variables identified	Up to 5
	4	Variables controlled	Up to 6
	5	Satisfactory implementation	Up to 6
	6	Treatment of results — appropriate	1
	7	Conclusions drawn correctly	Up to 3
	8	Conclusions related to initial hypothesis	Up to 6

NAME	1	2	3	4	5	6	7	8	COMMENTS	Strand levels (i)	(ii)	(iii)
R. Armstrong	✓	✗	✓	✓	✓	✓	✓	✗	Theory wrong — no evaluation in conclusion	2	4	2
S. Bell	✓	S	–	✗	✓	✓	✓	✓	Didn't control weight Good disc. results in terms of forces	5	4	5
T. Cook	✓	P	✗	✗	✓	✗	✗	✗	Used diff. height and weights Not able to draw conclusion	3	2	–
H. Davy	✓	–	✓	✗	✓	✓	✓	✗	Simple prediction no support theory cut wings recog. in conclusion	2	3	3
I. Edison									ABSENT	–		
D. Evans	✓	S	✗	✗	–	✓	✓	✓	Cut wings	4		

Figure 2 A class sheet for pupil records

Relevant information can be transferred to individual pupil records at the end of a topic or as the school policy dictates. It would seem sensible to maintain and keep such records on a departmental basis and to transfer them to a central school record at the end of the year. Decisions about record keeping should be a part of a whole school policy, but they need to be sympathetic to the problems faced by the teacher in the classroom.

4 Evaluation

The assessment information derived from whole classes or from groups within classes can be used by teachers for providing feedback on their own performance, both as developers of teaching schemes and as deliverers of the different parts of those schemes. The information will be more reliable, and so more useful, if it is complemented by the views of pupils as to their own reasons for their performance. If they have a clear understanding of the learning objectives and what is expected of them in terms of performance they can usually make an honest assessment of their own contribution to learning. Again, this can be incorporated in the development of the record of achievement.

Discussions about problems and shortcomings of the scheme of work will take place as a matter of course in well motivated departments. The head of department must ensure that time is made available in departmental meetings to ensure that these discussions have a positive outcome. They should form part of a continual review and refinement.

The school itself and the departments within it, will themselves be publicly evaluated via the GCSE results. It will be very unfair if this evaluation is based on 'raw' examination results. Assuming there is, across the whole country, a reasonably good correlation between pupils' performance at one key stage and the next, it would be very easy to give individual schools, departments (or even teachers) a score based on comparing what is *actually* achieved with those pupils with what, on the basis of national figures, a group of pupils with the same ability at the previous key stage 'should' have achieved.

Post-script – some words of caution.
The measuring instruments which we have at our disposal are imprecise, and what we are trying to measure is at times transient and liable to be affected by the measuring process. Models of the curriculum and its assessment are

human constructs just like the models which scientists have created to explain the universe. The difference is that the curriculum and its assessment rest on a very slim research base. Most of the foundations are the stuff of personal and political prejudice. Like most (or all?) human constructs these things are not immutable and they are not eternal. It would be at best unwise, and at worst wrong, to grant assessment the status of the measure of the quality of an education or a school, as some seem to want. There are plenty of examples of successful people who were 'failed' by the education system in the past because they 'failed' its assessment system. We need to maintain a sense of balance, building on those aspects of assessment which seem to be useful in assisting the progress of pupils and treating the others with the caution that they demand.

John Skevington is Head of Science at Sherburn High School. He is chair of ASE Assessment and Examinations Committee and a member of SEAC Science Committee

References

ASE (1992) *ASE policy present and future*, ASE.

British Association of Teachers of the Deaf, *The language of examinations*.

DES (1992), *Reporting pupils' achievements to parents*, Circular 5/92

Fairbrother R W (1988) *Methods of assessment*, Longman.

Industrial Society (1989) *Managing assessment*, The Industrial Society.

MEG (1992) *GCSE Coordinated Science syllabus: The Suffolk development*, Midland Examining Group.

NEAB, WJEC and ULEAC (1992) *GCSE Framework Common Themes*

Rowntree D (1987) *Assessing students: How shall we know them?*, Kogan Page.

SEAC (1992a) *KS3 school assessment folder Part 2*, School Examinations and Assessment Council.

SEAC (1992b) *GCSE Criteria for science*, School Examinations and Assessment Council.

University of Bristol (1992) *A whole-school assessment policy*, NFER-Nelson.

ULEAC (1992) *GCSE science syllabus: Graded Assessments in Science Project*, University of London Examinations and Assessment Council.

White R and Gunstone R (1992) *Probing understanding*, Falmer Press.

Managing and organising in science education

6 Management and Organisation of Science Teaching

Alan Boyle

As a Head of Science, or aspiring Head of Science, there are certain issues you will need to consider in the management of your department.

1 What Management is About

1.1 Getting other people to do their job

Secondary schools may vary in size but most are complex organisations that need managing. Management is about organising resources to achieve a satisfactory performance and getting other people to do their job. Consequently every teacher is a manager, if you consider the job or task to be learning, and the pupils as the people who complete that task. Somebody in the school should manage the staff, whose job it is to teach the pupils. In all schools you will find teachers with managerial responsibilities for other staff, as well as their own fundamental teaching role. These extra managerial responsibilities are usually rewarded with increased pay. Management structures in schools, which are commonly hierarchical, indicate lines of management from the head teacher, who has overall responsibility for the running of the school, to each member of staff.

The head of science in a secondary school usually has a large teaching load, together with responsibilities for organising and managing science teaching throughout the school. Consequently, time and energy have to be shared and this will lead to conflicting demands. Another potential source of conflict is trying to balance the specific interests of science education against the general interests of the whole school. The management and organisation of science teaching is both challenging and rewarding work when things go well: it may be frustrating at other times.

1.2 A management process

There is no guaranteed successful formula for organising and managing anything, let alone everything. There are, however, strategies that have been successful in certain situations. There are skills that managers use frequently and there are sensible ways to organise science teaching. To begin with, you

100

may find the following structure logical when going through the process of getting other people to do a job.

1. **Plan** Discussion and decisions are key features at this stage. Who is involved in discussion and decision making will depend on the manager's style and the situation. It may be a development plan for the next five years or a curriculum plan for the next five weeks. Whatever it is, the plan should clarify what has to be done and why. It will include setting measurable objectives.
2. **Organise** The resources required to do the job, both personnel and equipment, are considered at this stage. You will need to respond to questions such as: Who? When? Where? and What with? Constraints imposed by the school may lead to compromise. Written schedules and timetables are frequent outcomes.
3. **Implement** Even in well run departments, the most carefully organised plans cannot implement themselves. This stage in the process has three key managerial functions:
    *   motivation,
    *   communication,
    *   monitoring.

    Good managers learn how to motivate people continuously to do their job. The art of communication is ensuring that your message is received as you intend it to be. A monitoring system should record how things work out as well as checking that they go according to plan.
4. **Evaluate** The collection and analysis of evidence from different sources, including pupils and people outside the science department, will be useful in any evaluation. Prejudice should be avoided and reasoned judgement, supported by the available evidence, should aid future planning.

The process is cyclical and probably will begin with an evaluation of the way the job was done before. Think the process through in terms of the teacher's role, managing how pupils learn. Each stage in the process is related to managing the work of other teachers.

In carrying out this process successfully, the manager should also carry out another important function, i.e. develop the professional skills of staff. In this case the teacher and technician, as well as him or herself.

1.3 Management style

Successful management combines concern for people with the achievement of a shared task. These twin dimensions may be used to describe your management style. If concern for people and concern for task are represented on mutually perpendicular axes, (fig 1) four broad management styles are suggested by the polarised positions.

Superficially it may be assumed that the 'integrated' style, having high concern for both task and people, is the ideal management style. It is worth considering that while this style is frequently desirable, other styles are more suited to particular situations. When it is generally accepted that the achievement of the shared task is crucial then the autocratic (dedicated) style could be more appropriate. On the other hand, when achieving a task is less important than relationships with staff, then the more democratic (related) style may be favoured. Finally, if the task is relatively less important and it is unlikely to affect relationships, the bureaucratic (separated) style would work well. To be more effective, you may consider adopting appropriate management styles in different situations. This demands flexibility and resilience.

1.4 Science Teaching Quality Management System

Quality improvement assumes that no matter how good something is, it can always be improved. Continuous quality improvement is a process which may apply to science teaching as well as anything else. It requires an attitude

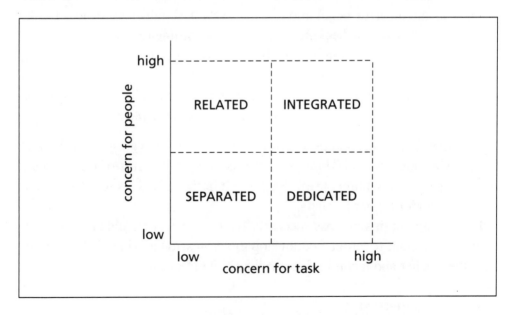

Figure 1 Management styles revealed by relative concerns for task and people

of never being satisfied with current success and a willingness to look for ways of improving, rather than maintaining, standards.

The Science Teaching Quality Management System described here is adapted from Virginia's Science Education Program Assessment Model. It has been designed to help a science department to develop its own quality improvement system. It attempts to identify pupils' and parents' expectations and perceptions, as well as providing quality assurance from other professionals. The suggested procedure shows how a science department could set up its own system.

1. *Decide who should be involved in the quality management process.* It is recommended that four groups participate in order to offer different perspectives. These groups are: students (S), science teachers (T), parents (P) and an external group (E), which may include senior managers, local inspectors or other teachers in the school. External in this sense means external to the science department.

2. *Prepare a list of criteria against which you want to evaluate the quality of science teaching in your school.* You may find it easier to start with a list of the general areas. For example, the Virginia system uses headings of:
   School support
   Accommodation
   Safety
   Resources
   Learning experiences
   Community involvement
   Evaluation
   Professional development
   It may be useful to consult the stated science department's aims. For example the following statements, adapted from the Virginia system, might be used for Learning Experiences. They are included to indicate the kind of style which might be used. (See the end of the chapter for further details)

Learning experiences
   32. *Student interest in science lessons is stimulated by the use of interesting and novel activities. (T, S)*
   33. *Science laboratories are kept in a neat and attractive manner. (T, S)*

34. *Students have ready access to books and equipment in science rooms/labs. (T, S)*
35. *Students' work is displayed in science rooms/labs. (T, S)*
36. *Artefacts are displayed at interest centres in science rooms/labs and changed frequently. (T, S)*
37. *Displays about science topics are shown in science rooms/ labs. (T, S)*
38. *People from the community are used to enrich science lessons. (T, S, P)*
39. *Students have practical work in science every week. (T, S)*
40. *Field trips are provided to enrich science teaching. (T, S, P)*
41. *Schemes of work for science provide all students with their entitlement according to the national curriculum. (T, E)*

3. *Allocate each criterion to at least two groups of participants for evaluation.* For example, in the above statements, criterion 41 "Schemes of work for science provide all students with their entitlement according to the national curriculum." has been labelled T and E. This indicates that the group of science teachers (T) and the external group (E) would both be used to evaluate the science department according to this criterion.

4. *Prepare four questionnaires, one for each group of participants, based on the criteria you have selected for the groups.* Some of your criteria may need to be adapted into more suitable phrases for students and parents. For example, criterion 40, 'Field trips are provided to enrich science teaching', could become 'My science class goes on field trips' for students and 'Science field trips are provided for my son/daughter' for parents.

5. Structure each question to allow respondents to indicate:
   (a) how desirable they consider each quality criterion to be,
   (b) how far science teaching in the school has achieved that criterion.
   Part of a sample questionnaire is shown in Table 1.

6. *Analysis of the questionnaire responses* will allow you to consider:
   (a) the priority each group gives to the quality criteria according to their desirability
   (b) to what extent each group considers the criteria have been achieved.
   This information should help you to judge how well you are doing and how you should prioritise any future action to make improvements, in terms of the quality criteria used in the questionnaires.

Table 1 Science Teaching Quality Management System

Sample questionnaire

We would like you to provide us with some information which will help us to improve the quality of science teaching in our school. Your views are important.

Each statement in this list has two rating scales with it, one in the left margin and one in the right margin.

The scale on the left is used to show how much you agree with the statement. Score it from 5 to 1 to indicate how desirable you think the statement is. 5 represents most desirable, 1 represents undesirable.

The scale on the right is used to show how far you think the statement has been achieved in your school. Score it from 5 to 1 to indicate what you think the level of achievement is. 5 represents fully achieved, 1 represents not achieved.

Start when you are sure you understand the two scales.

Complete the desirability scale for each statement first. Write the number from 5 – 1 in the space provided. When you have finished all statements, cover the left hand margin with a sheet of paper and complete the achievement scale for each statement.

Thank you for your help.

DESIRABILITY		ACHIEVEMENT
5 4 3 2 1		5 4 3 2 1
☐	1. My science course is neither too difficult nor too easy.	☐
☐	2. There is enough equipment for all students to do experiments in science lessons.	☐
☐	3. I often use computers in science lessons.	☐
☐	4. My science teacher uses interesting activities in our science lessons.	☐
☐	5. I do practical work every week in my science lessons.	☐
	etc	

2 Situations and Strategies

It may not be possible to describe every situation that a head of science will encounter while organising and managing science teaching. Here are four common situations that most heads of science are likely to experience:

A.   Managing innovation
B.   Delegating responsibilities
C.   Managing conflict
D.   Dealing with difficult people

Each situation is described as a case study to provide a context to provoke thought and discussion. As there are no 'correct' courses of action, some of the issues involved will be identified. Try to speculate about the likely outcome of certain responses. A few general points will be made for further consideration when similar situations are encountered in other schools.

2.1 Situation A: Managing innovation

Kathleen Kelly is head of science at St. Kevin's High School. The head teacher tells her that she wants the whole school curriculum for Key Stage 3 to be revised in line with proposals from the governors' Curriculum Committee. Kathleen is asked to produce an action plan in order to develop a new science curriculum in response to these proposals.

Michael Maloney is a teacher governor and also a science teacher. His views of the governors' proposals are expressed in comments such as: "What a load of rubbish."

Geraldine Reilly is a young science teacher whose ideas are considered to be progressive by her colleagues. She welcomes this opportunity to move the science curriculum forward.

Patrick O'Hara has been teaching science for 20 years. He leaves school promptly each day when lessons end. He thinks he would find it hard to change the way he teaches.

Mary McArtney and the other two science teachers accept the need for change but feel insecure after the continuous changes they have been forced to make since the introduction of the National Curriculum. They are highly committed teachers who dislike the idea of further change.

Imagine you are Kathleen Kelly. What would you do in this situation? Ideally you would need to know much more about St. Kevin's School of course. Think about the information provided and how you would attempt to deal with the head teacher's request.

Issues to think about

- How should Kathleen react to the governors' proposals and the head teacher's request?
- What do you think would be the best outcome from this situation?
- What strategies could Kathleen use to achieve this outcome?
- Consider the other science teachers and how they feel about the situation. Imagine yourself in their position. What would your reaction be to Kathleen's strategies? What would influence you most?

General points to consider

1. A problem-solving approach may be useful when managing change. If this is followed, the first crucial step is to define the problem clearly in order to find out precisely what should be changed.
2. A range of possible solutions should be prepared in outline. This calls for creative thought to produce imaginative ideas. Lateral thinking skills are useful but the ideas may be gathered from a wide range of sources, including pupils and parents as well as professionals.
3. After selecting the desired outcome, it will need to be worked out in detail together with the strategies for introducing it.
4. Before introducing change, think about the context in which the change will take place. List those people and things that will support the change and also those that will oppose it. Work out ways of using and increasing the support and also ways of reducing the resistance.
5. Plan how to monitor the change, what data you will need to evaluate it, who will collect those data, when the data will be analyzed to prepare an evaluation report and who will produce the report.

2.2 Situation B: Delegation of responsibilities

The science department at Oaklands school recently changed the science curriculum in years 9, 10 and 11 from a co-ordinated approach, moving classes between specialist teachers of biology, chemistry and physics, to an integrated approach. They are using two teachers to share the planning, organisation and teaching of a double award GCSE course, for two classes, in years 10 and 11, while Year 9 has been joined with the combined science course that already existed in years 7 and 8, where each class had a single teacher for science. This followed the introduction, post 16, of a modular science curriculum for A-level, AS level and BTEC qualifications.

Sally Humphreys, the head of science, is now considering how to reorganise the responsibilities within the department. The departmental structure that exists already is:

Teacher	Scale	Responsibility
Sally Humphreys	D	Head of science (and biology)
George Bateman	B	Head of chemistry
Candy Stone	B	Head of physics
Roger Wright (years 7 & 8)	A	Lower school science
Terry Stewart	–	Health education co-ordinator
Amanda Pritchard	–	Multicultural education co-ordinator
Glen Lord	–	Environmental education co-ordinator
Charanjit Singh	–	IT co-ordinator
Mike Aldridge	C	Year tutor
Stanley Jones		Deputy head

Issues to think about

- What are the advantages and disadvantages of changing the structure of responsibilities in the science department at Oaklands School?
- What would be your preferred management structure for the science curriculum at Oaklands School if you were to start without any teachers and you were able to employ them for the positions you specify?
- If you were Sally Humphreys, how could you adapt your preferred structure within the constraints of the existing responsibilities and allowances at Oaklands school?

General points to consider

1. All teachers should share some of the managerial responsibilities in a science department. The distribution of these responsibilities should be commensurate with the salary allowances.
2. A participative management style will encourage teachers to influence the work of the science department. The responsibility is to carry out the decisions agreed by the whole department.

3. The implicit purpose of sharing responsibilities should be to improve the quality of learning for the pupils. This becomes more accountable if the nature of these responsibilities is more explicitly linked to the quality of learning, rather than to the status of the person holding that responsibility. For example: *'Standards of pupils' work in Years 10 & 11'* could be a responsibility which is explicitly linked to the quality of learning, whereas *'Head of physics'* is more about the status of the person who holds that responsibility.
4. All responsibilities need not be fixed. Some issues will be important at some times but not at others. For example, a gender bias may be apparent in the curriculum following an analysis of pupils' performances. Investigating and correcting this could then become an important responsibility for a short time.
5. Some departmental responsibilities will probably be permanent, eg 'Assessment in Key Stage 3'. By rotating responsibilities among all teachers in the department, for fixed time periods, the teachers gain in their professional development.
6. All responsibilities should carry a written job description. The person who accepts the responsibility can discuss and negotiate this with the head of science.

2.3 Situation C: Managing conflict

Stuart MacDonald is head of science at Bonnybridge Academy. The science curriculum indicates that, when aggregating an overall grade for science, the relative weighting between the processes of science, and knowledge and understanding, should be 1:3 . The written scheme of work suggests that all knowledge and understanding should be gained, as far as possible, through an investigative approach. Schedules of experiments, investigations and practical work have been prepared for teachers to follow. Stuart has noticed that one of the science teachers, James Brown, does not carry out any practical work with his classes. James describes the experiments, the pupils copy diagrams from the text book and James gives them the results to use and interpret. The pupils' exercise books are neat, well-ordered and impressive. James's pupils perform well in their written examinations.

Stuart has approached James about the lack of practical work carried out by pupils in his lessons. James says that he thinks he is a good theoretical teacher and that this is far more important than practical work. James refuses to carry out the schedule of experiments in the scheme of work.

Issues to think about

- To what extent does James have autonomy in his classroom? If James' pupils perform well does it matter how that is achieved?
- What risks would Stuart run if he decided to ignore the situation?
- Imagine you were Stuart. What outcome would you want from this conflict?
- Imagine you were James. What outcome would you want?
- Create a possible solution that attempts to meet your views of both Stuart's and James's preferred outcomes, as far as possible.

General points to consider

1. Conflict offers opportunities to make progress. It should be viewed positively: it is considered to be essential in any healthy organisation. Continually avoiding conflict is a sign of weak management. Progressive organisations should be able to encourage conflict and deal with it creatively.

2. In order to solve conflict, it must be defined first. As it takes (at least) two people to produce conflict, both should be involved in defining the conflict from their own respective viewpoints.

3. There are several ways a manager can deal with conflict; these include:
    (a) ignore it
    (b) overpower your opponent
    (c) accept your opponent's point of view completely
    (d) seek a compromise by trading bits of one point of view for bits of the other
    (e) create a new solution which seeks both points of view

4. A manager should try to analyze the situation before deciding how to deal with the conflict. A simple analysis may help, based on the relative importance of the task to be achieved, or personal objective, and the relationship with the other person involved.

5. If the task, or personal objective, is very important, but the relationship is not, 3(b) is the most effective strategy. If the relationship is far more important than the task, 3(c) is best, in the interests of harmony. When both the task and the relationship are valued highly, 3(e) is the best course of action. When neither the task nor the relationship are important, 3(a) is a suitable response.

6. Creating a solution that achieves both your own goals and those of the other person is not easy. It requires imaginative thinking and takes time and energy. Consulting other people is a useful way to generate new ideas and possible solutions.

7. Avoid the temptation to choose 3(d) as a way of dealing with conflict. It will rarely satisfy either of the people involved and usually the conflict emerges again in a slightly different form.

8. Solving conflict creatively improves relationships by reducing tensions and produces better ways of achieving the task.

2.4 Situation D: Dealing with difficult people

Terry Carson is a head of science. He is concerned about the way that discussions in departmental meetings, or during school-based INSET, break down or avoid dealing with the issues involved. Here are two examples of how the discussions end up.

Terry: You have had an opportunity to study the new scheme of work for the magnetism module in Year 7; what do you think are the advantages and disadvantages of this new module?

Teacher A: I don't think we should be doing it this way. It has always worked perfectly well the way we did it before.

Teacher B: Yes, that's right. Remember when Tony changed the circuit board experiment a couple of years ago? Well, what a disaster! I could have told him it wouldn't work but he wouldn't listen.

Teacher A: No, they never do. Always trying out one fancy idea after another, I don't know where they get them from.

Teacher B: Well I blame it on the National Curriculum. Our kids used to love doing science until the National Curriculum came along and now look at them. Bored stiff every lesson. ....

...Ad nauseam.

Terry: What do you think about teaching new modules outside your subject specialism?

Teacher C: I would like to try teaching 'Food Chains'

Teacher D: You couldn't teach that.

Teacher C: Why not? I think I could.

Teacher D: Look, you've never taught it before. It's as simple as that.

Teacher C: I don't care. If I want to teach it I will, you can't stop me.

Teacher D: I'm not trying to stop you. I'm just telling you that you couldn't do it.
Teacher C: Amanda can teach it, so why can't I?
....and so on.

In both of these examples the conversations could easily have continued in the same manner for a considerable length of time.

Issues to think about

- What is likely to happen if Terry does not intervene?
- Think of some different intervention strategies that Terry could use.
- How could Terry use other teachers present?

General points to consider
'Transactional Analysis' may be useful to Terry to help him work out what is happening. This is a theory of social intercourse created by the psychiatrist Eric Berne, based on interactions between humans observed during psychotherapy groups. He noticed that certain sets of behaviour patterns exhibited by posture, voice, vocabulary and facial expression corresponded to three states of mind. He called these *Ego States* and they are described as reality, not theory, as they generate a coherent set of feelings that match the coherent set of behaviour patterns.

To simplify understanding, he called the three ego states: Parent (P), Adult (A), Child (C). In your P (Parent) ego state you resemble the parental figures in your life. This may be in a nurturing or controlling capacity. In your A (Adult) ego state you behave logically. You attempt to problem solve and consider things in a rational manner. In your C (Child) ego state you are sometimes playful, sometimes compliant and sometimes rebellious. All three ego states are normal and essential for the health of a human.

Successful transactions take place between people who are both in the same ego state. They are also successful between one person in the Parent ego state and another in the Child ego state. They are not successful when one is in the Adult ego state and the other is in either the Child or the Parent ego state.

In Terry's situation he is in his Adult ego state but he is unable to have successful transactions because his colleagues are in different ego states. Teachers A and B are both in their Parent ego state and will happily continue their transactions for some time. Teacher D is also in the Parent ego state and Teacher C responds by moving into the Child ego state. These transactions will also continue successfully.

Terry will only be able to have a successful transaction with any of these teachers if they change to their Adult ego state or if he changes to his Parent or Child. The greatest danger for Terry is that he engages in their transactions, as he will then find it impossible to achieve his intention, which is to have A – A transactions. He can attempt to intervene by using a set of prepared cues which will stop the P – P or P – C transactions and invite his colleagues to respond by changing to their Adult ego state eg "Is it true that . . ."; "Does it apply to . . ."; "Is that appropriate . . ."; "What is the evidence for . . ."; "Does anybody else share that opinion?"

It is important to remember that the Parent and Child ego states are just as important as the Adult ego state. It is likely that during a departmental meeting, the Adult is the most successful ego state for dealing with the work. In different social situations the other ego states are more successful. The following suggestions may help you to stay in your Adult ego state.

- Be sensitive to P and C signals in others (facial expression, tone of voice, vocabulary, physical gestures)
- Know your own P and C feelings and signals
- Change to your A by processing data (count slowly to 10, make a conscious change)
- Turn the situation into a problem solving activity
- The strength of the A is in restraint. If in doubt, don't make a transaction.

3 Skills to Develop

3.1 Planning and running meetings

Apart from informal discussions during coffee and lunch breaks, a science department needs to meet regularly to discuss policy, express opinions, share information, consult and take decisions. As most meetings take place in teachers' own time, it is wise to call meetings only when it is essential. Most information sharing can be done on paper. Discussions about policy, consultation and decision taking require groups to meet. If they are chaired well, science department meetings make people feel involved and build teams.

All meetings should have a simple agenda circulated at least three days before the meeting, together with the minutes from the previous meeting. When preparing the agenda take care with the order of the main items. It is wise to include items which require lengthy discussion early in the meeting, or even as a single main item.

Northwood Community Comprehensive School
Science department meeting
Tuesday 8th December 1992, 4.00pm – 5.00pm, Laboratory 1

Agenda

1. Apologies for absence
2. Minutes from the last meeting
3. Matters arising from the minutes
4. Consultation on Curriculum Board proposals for Year 10, September 1993
5. Review of Year 7 scheme of work
6. Arrangements for mock A-level examinations
7. Decision about residential field trip, June 1993
8. Any other business

Figure 2 Agenda for a science department meeting

The responsibilities of the chairperson include:

- starting and finishing the meetings on time
- arranging and following the agenda
- allowing sufficient time for each item
- ensuring that everyone can express their views.

Notes should be kept by another member of the meeting and the minutes prepared in consultation with the chairperson. Keeping notes at the meeting is usually shared by members of the department on a rota.

Useful tips

1. Arrange the meeting around a table in a pleasant room.
2. Be prepared by marking the previous minutes with matters to be raised.
3. Avoid re-living the last meeting, by treating 'matters arising' briefly. Any important follow-up should be a main item in the agenda.
4. Discussion documents about important items should be circulated with the agenda.

5. Members of the department may introduce items but the chair should retain control of the discussion.
6. When a lot of people want to speak, it is useful to keep a running order and, between speakers, to state what it is.
7. Humour and a sense of perspective are valuable during heated discussions.
8. Use 'lack of time' to shorten long-winded contributions.
9. Dispose of the irrelevant contributions politely but firmly.
10. Signal the end of a debate a few minutes beforehand.
11. All actions agreed during the meeting should be clearly noted and read out again at the end to make sure that those responsible for them are aware that they will be minuted.

3.2 Appointing new staff

It is unlikely that a head of science will have total responsibility for the appointment of new teachers. It is important that a head of department is involved in short-listing and interviewing applicants for posts within the department. In most schools this is common practice. For many, this will be an unfamiliar experience. The head of science will be expected to give professional advice about the candidates' abilities related to science education and how their qualities match those of existing teachers in the department. Unless the head of science is also a teacher-governor, he or she is not expected to take part in the decision.

Job descriptions and person specifications should be created for all appointments. They may already exist from previous appointments and are an essential guide to choosing the most suitable candidate. An example is shown in Figure 3 for an assistant teacher. This would be added to for posts of responsibility. Both the short-listing process and the interview questions should be related to criteria derived from the job description and person specification.

During the interview, the head of science may be expected to ask questions that relate specifically to science education. These can be derived from contemporary issues general to science teaching in all secondary schools and specific issues important to the current state of development of science teaching in the school.

Useful tips
1. Try to form open-ended questions that allow the applicants to express their own opinions.
2. Do not ask leading questions to which you think you have the right answer.

3. Avoid long rambling questions and multiple questions with several parts.
4. Prepare the questions before the interview and try them out on other colleagues.
5. Be ready to stop a candidate who rambles off the subject when giving an answer.

Figure 3 Sample job description and person specification

Westvale Community Comprehensive School
Science teacher job description

Responsible to Head of Science Department

1. To teach science throughout the school as directed by the Head of Science.
2. To plan and prepare lessons according to the agreed scheme of work used by the science department.
3. To assess and record the progress of pupils according to the science department assessment policy.
4. To participate in science curriculum development with other members of the department.
5. To assist the head of department in the work of the department.
6. To carry out the duties of a form and subject teacher as described in the general specification.
7. To fulfil the school's homework policy.

Science teacher person specification

1. Qualified teacher
2. A higher education qualification in a science subject
3. Enthusiasm to teach science
4. An understanding of science in the National Curriculum
5. An awareness of how pupils learn science
6. Ability to relate to young people
7. Willingness to take part in extra-curricular activities

3.3 Organising school-based INSET

The professional development of staff is a key managerial responsibility. As a middle manager in a secondary school, a head of science may be responsible for planning and organising training days that are related to the science department's expressed needs.

The professional development process is cyclical. It has no beginning and end as such, but it may be useful to start with the identification of needs. Teachers' training needs vary in three broad areas: (i) their personal needs, (ii) the department's needs, (iii) whole school needs.

The personal needs relate to professional competencies and career aspirations. They may be identified through appraisal (see p 118). The department's needs should be curriculum led and respond to an evaluation of science teaching and the standard of pupils' work. Whole school needs will be indicated in the school development plan. Having identified a range of training needs, some prioritising is needed to prepare a programme for professional development.

Part of the professional development programme will be school based and those aspects that relate specifically to science may be the responsibility of the head of science. When organising the programme, pay attention to each of the following:

- intended outcomes
- resources available
- variety of learning activities
- look for progression
- seek guidance and support
- evaluation

During the training itself, arrange to look after the providers as well as the staff being trained. The trainers may be colleagues, teachers from other schools, advisers, consultants or others.

Evaluating a professional development programme is a long term process. Immediate feedback is helpful, but the real test of its success can only be judged by the effect in the classroom. A useful way to evaluate INSET is to judge it against its intended impact. The following levels of impact may be used:

- Raising awareness of a new topic or technique
- Developing an understanding of the principles involved
- Application of the principles with some feedback
- Introduction of the principles into normal practice

3.4 Appraisal

The identification of a person's professional strengths and weaknesses may be achieved through an appraisal system. If we accept that nobody is perfect and that gradual continuous improvement is desirable, appraisal should identify opportunities for professional development. To be successful, it must be supported by an entitlement to negotiated training provision.

In most secondary schools the head of science will be involved as an appraiser as well as being appraised.

Appraisal systems usually include some, or all, of the following:

- self evaluation by the teacher
- classroom observations by the appraiser
- formal discussion, with structured preparation on both sides, leading to an agreed statement
- acknowledgment of strengths
- identification of opportunities for improvement
- written record of the process signed by both parties

When making classroom observations a head of science should know beforehand what to look for. This usually involves the production of an observation schedule containing a list of criteria. The schedule should:

- be linked to professional competencies
- contain general criteria related to teaching any subject
- contain specific criteria related to science teaching
- be shared with teachers before lessons are observed
- have some input from teachers in its design

Allow sufficient time for the formal discussion. It should take place in privacy without the possibility of any interruptions. Professional integrity should ensure confidentiality and the appraiser needs to be sensitive to the feelings of the teacher. The discussion should focus on agreed professional competencies that the teacher has evaluated personally and the appraiser has observed during the lessons visited. Personal qualities should not be discussed. The outcome should make the teacher feel valued and motivated to improve his or her professional skills.

4 Getting Organised

The day to day running of a science department requires a high degree of organisational skill. Some of this work will be delegated to teachers within the department, but the head of science is responsible for ensuring that it works.

4.1 The Curriculum

Planning and organising the curriculum follows a simple and logical sequence, from the school's aims to activities in each individual lesson (figure 4).

The head of science should manage the whole sequence, with specific responsibility for developing a rationale and designing schemes of work, although some of this could be delegated. Individual teachers are then responsible for preparing their own lesson plans.

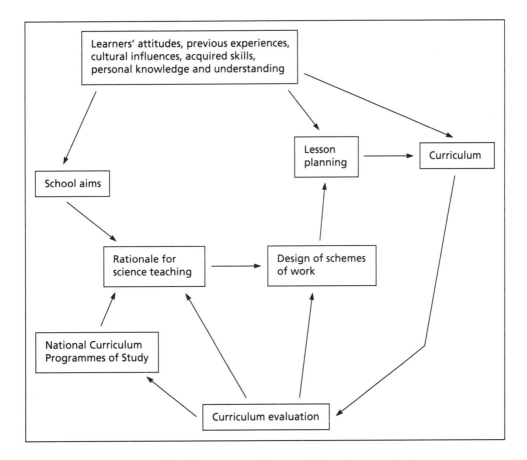

Figure 4 A sequence for planning and organising the curriculum

4.1.1 Design

Curriculum design is an organisational task which plans situations where the aims and intended learning outcomes may be achieved. The rationale will provide an indication of the way that the curriculum will be arranged. A scheme of work or teaching schedule shows how the curriculum is designed. This is detailed work and requires knowledge of the resources available to support the planned activities. An agreed format for all schemes of work in a school helps any teachers working in new areas. Whatever the format, a scheme of work should include the following:

- an overview showing the sequence of learning activities
- reference to the intended learning outcomes for each activity
- brief details of the activity
- lists of the resources available (equipment, texts, work-sheets)
- ideas for dealing with pupils of different abilities
- details of risks or hazards associated with the activity
- how to assess pupils who have completed the activity
- some indication of typical pupil outcomes
- length of time needed for the activity
- suggested opportunities for further development

4.1.2 Resources

The management process outlined on page 101 shows that changing plans into reality requires careful consideration of resources. It is important that the resources needed are first considered without any constraints, so that the intended goals are achieved. This is difficult to do, as the stated intentions are sometimes influenced by a knowledge of existing resources. If continual improvement is desired, then curriculum planners should be more open-minded and consider what would be ideal. An audit of existing resources can then be matched to the ideal requirements. This will identify areas where resources can be improved and help a department to consider how to be most effective, efficient and economic.

4.1.3 Staffing

(a) Teaching staff

When it comes to staffing the science curriculum a head of science needs to consider the following:

- the number of teaching groups
- the number of lessons per group
- the amount of non-contact time for science teachers

A simple calculation will then determine the number of teachers needed to staff the science curriculum. It is useful to begin by first considering the ideal.

The number of teaching groups for science will depend on the number of pupils on roll and the size of each group. ASE policy states that practical work in science, for pupils aged 5–16, should be taught to groups of no more than 20 pupils per teacher, for safety reasons. If the curriculum is planned so that science is taught through an investigative approach, this will have implications for the number of teaching groups. Another important factor is the number of pupils with special educational needs. Extra teachers may be timetabled to provide support for these pupils in large groups, or much smaller teaching groups may be formed. These decisions would be influenced by the school's policy for pupils with special needs.

The number of science lessons per week for each group obviously depends on the length of each lesson. When Science in the National Curriculum was planned, it was assumed that it would occupy 10% of curriculum time in years 7 and 8, 15% in year 9 and 20% in years 10 and 11 for those pupils following the double science course (12% for pupils following the single science course).

These percentages can then be applied to the DES (now the DFE) recommended minimum number of hours needed per week to teach the national curriculum (23 hours for pupils aged 11-14 and 25 hours for pupils aged 14-16, DES Circular 7/90) to calculate the following minimum total weekly lesson times recommended to teach Science in the National Curriculum:

Years 7 and 8:   2 hours 21 minutes
Year 9:   3 hours 32 minutes
Years 10 and 11:   5 hours (double science)
                    3 hours 8 minutes (single science)
(NB These recommendations are not statutory.)

The average amount of non-contact time for science teachers in a school is calculated from the school's average contact ratio.

$$\text{average contact ratio} = \frac{\text{no. of periods taught}}{\text{no. of teachers} \times \text{no. of periods}}$$

Applying these parameters to an imaginary situation of a school with 180 pupils in each year, teaching double science to all pupils in years 10 and 11, with forty 35 minute periods per week and an average contact ratio of 0.875, gives the following data:

Contact time   Year 7:   36 teacher periods
             Year 8 :   36 teacher periods
             Year 9 :   54 teacher periods
             Year 10:   72 teacher periods
             Year 11:   72 teacher periods
             TOTAL: 270 teacher periods
Non-contact time =   39 teacher periods

This gives a desired staffing level of 7.7 full-time teachers.

It is at this stage that desirability has to be matched with expediency. The size of teaching groups may be increased, the number of lessons taught may be reduced, or the contact ratio may be increased, to reduce the number of teacher periods required to staff the curriculum. These judgements are very important as effectiveness and efficiency are often traded for each other.

(b) Non-teaching staff

The amount of practical work in science lessons, the sophistication of much of the equipment and the complex nature of the tasks involved, mean that secondary schools need the support of science technicians. The essential level of support required will depend to some extent on the curriculum and the lay-out of the school laboratories. ASE policy indicates a crude formula that may be used to calculate the minimum amount of technical support needed:

Total no. of technician hours per week = total no. of science teaching hours × 0.85

Using the same example above for staffing the curriculum, there are 270 science teaching periods. This means that the school would need a minimum of 157.5 technician hours per week which is equivalent to four full-time technicians.

4.1.4 Materials

Efficient organisation of equipment, consumables and textbooks will allow science teachers to use careful stock control and avoid wastage of the limited budget available for resources. Efficiency and economy are in the department's own interest as they allow a more effective use of the budget. The use of a data base will enable the department to store all this information and search it at high speeds. Commercial programs, specifically designed for science departments, are available from software publishers. Whether electronic or paper records are being used, lists will be needed for:

- items of equipment
- consumable stock
- textbooks

Keeping accurate and up to date records of these may be tedious and time consuming but it will benefit the departmental budget.

(a) Equipment
An inventory of items of equipment over £50 would probably be required, for an auditor to check. The inventory should contain the following information:

- item
- serial number
- supplier
- catalogue number
- date of purchase
- price
- estimated useful life
- annual depreciation
- location
- last annual check

The total annual depreciation of major items of equipment may be calculated from this inventory. This should be adjusted annually by the average rate of inflation over the previous twelve months to give the department some idea of this aspect of its budget requirements. Obviously it would not be possible to replace all major items of equipment at once, but this information would help a department to plan for replacements over a number of years.

(b) Consumables
Chemicals, glassware, electrical equipment, notebooks, paper and work-sheets are all consumables. Separate lists could be compiled for different groups of consumables. Close checks on levels of stock are needed to enable the department to apply stock controls when necessary and also to predict the annual cost of consumables. A record of consumables in stock should show:

- consumable item
- location of stock
- date
- quantity in stock
- quantity on order (with order no.)

- supplier
- catalogue number
- cost
- date received (add to quantity in stock)
- estimated annual consumption

(c) Textbooks

While individual teachers should keep records of books loaned to pupils, the department should keep a central record of all numbered textbooks with pupils' names. The central record should indicate the condition of each book on the date of an annual check.

Loss of textbooks is likely unless all staff agree and enforce a strict policy. Some schools require pupils to have clearance forms checked before they leave. If accurate central records are kept, it is always possible to write to parents and request the return of missing books. When doing so it may help to say, "According to our records, your ......". This avoids embarrassment if the pupil has returned the book to another teacher.

4.1.5 Finance

The proportion of the school's budget allocated to science will include salaries of teaching and non-teaching staff as well as the amount spent on equipment and materials. As discussed above, the staffing costs are not fixed and are affected by group sizes, science curriculum time and staff contact ratio. All of these are variable between schools and also within the same school. The cost of replacing equipment, consumable materials and books will also vary according to the amount of practical work in the curriculum and the number of pupils in the school. In 1990 The Royal Society published a report, (Royal Society 1990) in which the running costs for a science department, teaching science as described in the National Curriculum, were estimated. The report concluded that, in 1990, a sum of £8.86 per pupil per annum was required.

Schools operate a variety of systems for allocating money to departments. There are dangers with closed systems. They involve secrecy, may indicate an unhealthy organisation and, in the worst scenario, are more open to the possibility of corruption, as the system relies entirely on the integrity of the benefactor, whose integrity, some would argue, is suspect for operating the system in the first place! An open system allows the whole school to see how the budget is allocated and should require departments to account for how it is actually spent. The two most common open systems used are formula funding and open bids.

(a) Formula funding

A formula used in schools should include:

- the number of pupils studying the subject (N).
- a weighting for the age of the pupils (Y1) for Y7-9, (Y2) for Y10 & 11, (Y3) for Y12 & 13
- the number of periods on the timetable (p)
- a weighting for subject allowances: (e) for equipment, (t) for text-books, (c) for consumables.

The age weightings could be the same as those used to determine the school's own delegated budget. The subject weightings are more con-tentious, but attempt to indicate the departmental needs. For example, equip-ment weightings (e) may be 3 for science and technology; 2 for music and PE; 1 for all others. Textbook weightings (t) may be 3 for English; 2 for mathematics, science, history, geography, languages, and RE; 1 for art and music; 0 for PE. Consumable weightings would be high for science, technol-ogy, and art.

The total allowances should be calculated for each department across each year of the school according to the formula:

Departmental allowance, $A = E + T + C$ where $E = (N \times Y \times p \times e)$
(for each year group), $T = (N \times Y \times t)$
                                                                                                 $C = (N \times Y \times p \times c)$

Therefore $A = NY(pe + t + pc)$

Each department's allowances should then be totalled together with the total of all the allowances for all departments. The cash can then be distributed according to:

$$\text{Department Share} = \frac{\text{Department's total allowance} \times \text{Budget}}{\text{Total allowances for all departments}}$$

(b) Open bids

When bidding for budget shares the science department needs to have accu-rate estimates of annual expenditures on consumables, equipment deprecia-tion and textbook costs. Departments should be required to account publicly, at the end of each financial year, for the way that their allocation was spent.

An auditor may then identify inflated bids and adjustments can be made in the following financial year. A bidding system should also include extraordinary bids for the introduction of new courses.

Value for money is an important measure when schools have limited budgets at their disposal. To get some feel for this concept, a department should complete a cost-effectiveness audit for particular courses. This would include estimates of the proportion of staff salaries used, the cost of materials purchased specifically for the course and an estimate of what proportion of the department's equipment and consumables costs are attributable to the course. The total cost may then be compared to the course outcomes, both intended and actual. This information would be part of the department's evaluation of its work.

4.2 Accommodation

The Programmes of Study for science in the National Curriculum indicate the requirements for practical work. They require a considerable amount of investigation work that pupils need to carry out themselves. In addition to the usual science experiments, pupils are also required to carry out data searches and retrieve information from a variety of sources. All this has implications for the type of accommodation needed for science teaching.

Future designs for science laboratories need to provide the following:

- Enough flexibility for pupils to have access to computer terminals with a modem, reference books, magazines, video and the usual services for gas, water and electricity.
- A layout that allows different arrangements of the furniture for individual study, small group work and whole class teaching.
- Storage space to permit safe and easy access by pupils to basic equipment needed for investigations.
- Safety features which include fire extinguishers, fire blankets, sand buckets, isolation switches for gas and electricity and first aid kits.
- Adequate lighting and the ability to darken the laboratory.

The size of a laboratory depends largely on the class size. While ASE policy supports group sizes with a maximum of 20 for practical work, it is most likely that schools will need laboratories large enough to accommodate up to 30 pupils. A space of 75 square metres would be the minimum size for a group of 30. The number of laboratories required can be determined by the curriculum demands. Servicing laboratories means that calculations should not assume that all the laboratories are available every lesson. A reasonable guide is to work with 85% of the total number of laboratory lesson spaces. In the example used for staffing, (page 122), 270 science lessons were needed in a six form entry, 11-16, school. If this is 85% of the total science laboratory space available, then 318 lesson spaces will be needed in laboratories, so 8 laboratories are required. It must be pointed out that this figure is derived from the desired staffing ratio and a small teaching group size, with the assumption that all lessons will be taught in a laboratory. As a more general rule $(n + 1)$ indicates the absolute minimum number of laboratories needed for an 11–16 school, where n is the number of forms of entry.

For 11–18 schools the general rule is $(n + 2)$ as a minimum. With the increased demands on practical work made by the National Curriculum and the desirability for small group sizes, it would seem safer to work on $(n + 2)$ and $(n + 4)$ for 11–16 and 11–18 schools respectively.

The Science Teaching Quality Management System

If you are interested in obtaining a complete Science Teaching Quality Management System that includes: full instructions, prepared questionnaires and a software package to help you to analyze the results, contact Alan Boyle, c/o ASE headquarters.

Alan Boyle was a head of department for 12 of the 20 years he taught science. He joined Knowsley LEA as senior inspector after being a Regional Project Officer for the SSCR. He has been a member of the ASE Policy Planning Group.

References: Further reading for aspects of this chapter

Management
    Kanter R M (1989) *When giants learn to dance,* London, Simon & Schuster.

Management style
    Reddin W J (1970) *Managerial effectiveness,* New York, McGraw Hill.

Quality management
    Neave H R (1990) *The Deming dimension,* Knoxville Tennesee, SPC Press.
    Exline J D & Tonelson S W (1987) *Virginia's science education program assessment model,* Washington, NSTA.

Innovation

Hull R & Adams H (1981) *Decisions in the science department: organisation and curriculum*, ASE.

Conflict

Pascale R T (1990) *Managing on the edge*, London, Viking.

Transactional analysis

Berne E (1970) *The Games People Play*, London. Penguin.

Meetings

ASE (1991) *Should we go on meeting like this?*, ASE.

Appointments

Paisey A (1985) *Jobs in schools*, London, Heinemann.

INSET

Oldroyd D & Hall V (1988) *Managing professional development and INSET*, unpublished MSc dissertation, University of Bristol.

Resources

Royal Society (1990) *The equipment resources required for teaching balanced science*, The Royal Society, London.

Finance

ASE (1991) *Getting to grips with LMS*. ASE.

Accommodation

ASE (1989) *Building for science: a laboratory design guide*, ASE.

7 Safety in Secondary School Science

Dr Peter Borrows

Peter Borrows is Chair of the ASE Laboratory Safeguards Subcommittee. The views expressed in this chapter are his own or those of this committee and not necessarily those of his employer.

1 Science is Safe

1.1 Safety in school laboratories

This Chapter can only give a brief overview of issues relating to safety in school laboratories. It is written especially to give a few pointers to those taking up science teaching, or management posts within science departments. However, all science teachers will need to read more detailed publications, such as reference [1], and need access to a range of reference sources such as references [2] to [9]. In addition, teachers need to be aware that safety advice can, and does, change. They have a professional responsibility to keep up to date, by watching out for safety advice in ASE journals, especially the Safety Notes in *Education in Science,* and the longer articles in *School Science Review.*

1.2 Accidents in schools

School science is very safe (see Table 1). The safest place a child can be is at school – much safer than being at home, much safer than travelling between home and school. In school, the laboratory is just about the safest place – much safer than the sports field, the corridors, or the ordinary classroom.

1.2.1 Accidents in laboratories

When you analyze the accidents which do take place in school laboratories (see Table 2) most of them are quite trivial for example – pupils burning their fingers because they attempted to pick up a hot tripod: burnt fingers in the laboratory are as much a part of growing up as grazed knees from playground activities.

However, a small number of incidents do occur in which pupils or their teachers are seriously injured, or which result in long-term health problems (see section 9). Awareness of the causes of such incidents will do much to eliminate them, and that is the responsibility of every science teacher.

Table 1. Percentage of fatal and major injuries to pupils in England, 1986/90 occurring in different parts of school (source: Health & Safety Executive)

Location	%
Playing fields & playgrounds	32.8
Gymnasia	25.1
Play areas	21.3
Corridors, etc	7.0
Classrooms	5.7
Extra-mural activities (not sport)	1.4
Toilets, etc	1.2
Laboratories	0.7
Craft workshops	0.4
Swimming pools	0.4
Other	4.4

Table 2 : Percentage of accidents reported from school laboratories (source: CLEAPSE School Science Service)

Type of accident	%
Chemicals in the eye	22.8
Chemicals elsewhere on body	20.6
Cuts	20.3
Burns from flames/hot objects	14.5
Dropping/falling/slipping/knocking/lifting	7.3
Chemicals in mouth	4.1
Inhalation	3.7
Animal bites	2.8
Explosions	1.5
Fainting	1.5
Electric shock	0.6

2 The Law and the Science Teacher

2.1 The Health and Safety at Work Act

Many science teachers at some stage express concern that they may be facing legal action, often as a result of something that they do not perceive as their fault. In practice, the risks are almost negligible.

The main legislation covering safety in schools is the "Health and Safety at Work etc Act 1974". This act is enabling legislation: that is to say it permits parliament to introduce new regulations, of which the COSHH ones (see section 8 below) have probably had the greatest impact on schools. It imposes a duty on the employer:

> *... to ensure as far as is reasonably practicable, the health, safety, and welfare at work of all his employees...*

There is also a duty

> *...on every employee while at work, to take reasonable care for the health and safety of himself and other persons who may be affected by his acts or omissions at work...*

Thus the main thrust of the act is to protect employees, including teachers and technicians, although most teachers would see their primary duty as being towards pupils. It is worth stressing employees' mutual responsibility: teachers need to be aware of the safety implications of what they are asking technicians to do; technicians have a duty to warn teachers if they consider they are planning to do something dangerous.

A successful prosecution under the act could result in a prison sentence of up to two years, and/or a fine, recently increased to a maximum of £10 000. However, it is the employer who is most at risk of prosecution: at the time of writing this chapter, in the 18 years since the act became law, only one science teacher has been prosecuted: he admitted that he knew the situation to be dangerous, failed to use safety equipment that was readily available, and disregarded quite explicit instructions from his employer, the LEA.

Employees are obliged to co-operate with their employer on health and safety matters. This means following whatever local rules their employer lays down, irrespective, for example, of what may be written in any safety publication, including this chapter, although it is considered that the advice here represents good, safe practice.

2.2 Civil law

As an alternative to prosecution under the criminal law, a teacher could be sued in the civil courts: an injured child (or a parent on their behalf), could sue for damages. In practice, the employer would be more likely to be sued, because the potential for damages is much greater.

2.3 ASE insurance

Under its scheme of insurance, ASE members are automatically covered against any civil action taken against them in the courts for the death of, or injury to, any person and loss of, or damage to, property either happening, or caused, during the performance of members' professional duties. This would also cover legal costs during any prosecution – but not the payment of any fines.

3 The Management of Safety

3.1 Employer's policy statements

Under the "Health and Safety at Work etc Act 1974", each employer must have a safety policy. In this context, the employer is the body with whom a teacher's contract of employment is made. For the majority of teachers, even under Local Management of Schools (LMS), this will be the Local Authority (LEA in England and Wales, REA in Scotland or ELB in Northern Ireland). For independent and grant maintained (GM) schools and City Technology Colleges (CTC's) the employer is the governing body. Most education authorities will in turn require that each school has its own safety policy. Science teachers should familiarise themselves with such policies.

3.2 The need for science department policies

Within schools, and particularly within science departments, in the past, safety has tended to be managed in an informal way: a chat with the new teacher when they asked for help; an occasional check of water levels in the jar of white phosphorus; purchase of new safety spectacles when everybody has been moaning about them for months. Such an approach is no longer acceptable. Changing legislation, changing attitudes towards safety, and changing responsibilities within schools mean that science departments or faculties need to have a written safety policy (see Section 4). Each department is unique, and needs to write its own policy in the light of its own circumstances, although these policies must of course be compatible with those of the school, the local authority, and any national requirements.

3.3 ASE INSET services

In order to help Heads of Science Departments, or those aspiring to such posts, the ASE Laboratory Safeguards Subcommittee has developed a one-day INSET course on the Management of Safety. Those interested in taking part in such a course should contact ASE INSET Services (via ASE HQ) to find out dates and venues of forthcoming courses. One of the main activities on the course involves looking at a possible structure for a departmental safety policy.

4 Departmental Safety Policies

4.1 Purpose

A safety policy is *not* a statement of all the ways in which practical science activities are to be carried out safely. Such information is already available in many publications [1, 2, 3, 4, 5, 9]. Rather, it defines procedures and areas of responsibility, in order to promote safe working for staff and pupils alike [8]. Any policy needs to be reviewed and revised regularly. Individual members of the department will need copies of it, perhaps as a part of a departmental handbook. It may be thought desirable to ask staff to sign a statement that they have received it. Much of the content is information about how the department works, which a teacher new to the department needs, and can reasonably ask for, from their Head of Department or mentor. The existence of a policy would be evidence of sound safety planning, and might provide a Head of Department with a good defence in the unlikely event of a prosecution or civil action.

4.2 Content

A safety policy document might contain the following sections:

- Introduction
- Specific responsibilities of particular staff
- General responsibilities of all staff
- Procedures
- Regular checks
- Pupils
- List of safety resources

4.2.1 Responsibilities of staff

The policy should lay down a chain of communication within the department. It will need to define who is responsible for what, naming names. The Head of Department has overall responsibility, but parts of the job are likely to be delegated. For example, who is responsible for :

- co-ordinating the arrangements for risk assessments under the COSHH Regulations?
- checking the storage and labelling of chemicals?
- checking that radioactive sources are logged in and out? (ie who is the Radiation Protection Supervisor?)
- checking first aid boxes?
- identifying safety issues in, for example, the Year 8 course?
- training staff (teachers and technicians) in the use of safety equipment?
- the induction of new teachers and technicians?
- informing responsible authorities when faults are found in safety equipment such as fume cupboards?
- monitoring the effectiveness of the departmental policy?
- monitoring the effectiveness of teaching safety attitudes to pupils? etc.

Such responsibilities need to be explicitly stated in the safety policy. Most science teachers would acknowledge that certain individuals have particular responsibilities by virtue of the posts they occupy (eg in their role as Head of Chemistry): it can sometimes be forgotten that all teachers and technicians have certain general ones. These expectations, (in effect, rules for teachers and technicians), need to be spelled out (see section 5).

4.2.2 Procedures

(a) Procedures for the day to day running of the department

The departmental safety policy should lay down procedures for dealing with the day to day running of the department, for example:

- how laboratories are to be left at the end of a lesson
- duties of the last person to leave at night
- emergency procedures
- what forms need to be completed in the event of an accident
- how COSHH Risk Assessments are to be carried out
- how staff are to be alerted to identified hazards in particular courses

- what procedures for risk assessment are to be adopted in open-ended investigations or projects
- how new staff, including technicians, students on teaching practice, probationary, articled, licensed or authorised teachers, or indeed those just new to the school, are to be inducted into its procedures
- how safety matters get on to the agenda of departmental meetings
- how pupils are to be trained in safety awareness
- what support is given to supply teachers covering lessons
- in what circumstances non-scientists may teach in laboratories
- how visitors, eg support staff or parents, are dealt with.

In addition, there may be rules governing the use of particular items, eg Local Rules for the use of closed/open radioactive sources, or a requirement that technicians put out the relevant CLEAPSS *Hazcard* [3] when preparing lessons involving the use of hazardous chemicals.

(b) Emergency procedures

Any departmental policy should address emergency procedures – evacuation routes, location of gas and electricity cut-offs, who to phone for advice, etc.

Under the First Aid Regulations only those who are trained in first aid should administer first aid. However, there are certain immediate remedial measures which all science staff should be prepared to carry out whilst waiting for first aid or professional medical treatment. For example, in the case of heat or chemical burns, or chemical splashes, washing the affected area for some time with water can reduce the risk of serious damage. The safety policy should spell out what is expected of staff.

All science departments should have spillage kits kept ready for use in emergencies. See Appendix F of reference [1] for guidance about the content of such kits.

4.3 Regular checks

Regular checks are required on a range of things found in science departments, for example:

- fume cupboards
- portable electrical appliances
- autoclaves, pressure cookers & steam engines
- radioactive sources for possible leakage
- chemicals likely to deteriorate

- eye protection
- first aid boxes
- fire extinguishers
  etc

Whilst there is a statutory requirement for checking some of the above items, for others it is just good safe practice. Some checks may be carried out by the employer, others are clearly the responsibility of the department. A science department policy should specify :

- where the responsibility lies
- the frequency and/or timing of such checks
- the procedures or criteria to be used
- how records of the check are to be maintained

Checklists of jobs to be done by the last member of staff to leave at night might be included, together with checklists of those jobs done on a weekly, termly or annual basis.

4.4 Safety rules for pupils

A departmental safety policy will include a copy of the safety rules for pupils, together with a statement of the policy on how safety is to be taught to pupils. It may also be appropriate to include a summary of the disciplinary procedures.

4.5 Safety resource file

The department should keep a central file of safety resources, known to, and easily accessible by, all members of the department. Included in this file will be any *General Risk Assessments* adopted by the employer, eg [1, 2, 3, 4, 5, 9], and any *Special Risk Assessments* obtained (see section 8). Relevant *Administrative Memoranda* from the DES and DFE, publications from the Health and Safety Executive, or circulars from the local authority should also be kept there for reference.

5 Responsibilities of Teachers and Technicians

It is the responsibility of all staff (teaching and technical) within a department to ensure that safe working practices are adopted. This includes :

- designing practical work in such a way as to avoid or minimise hazards
- seeking the safest practicable way of achieving desirable educational objectives

- emphasising individual's responsibilities to implement safety policies, procedures and precautions
- being aware of the uses of various types of safety equipment.

Staff have a responsibility towards each other, towards pupils, and also towards others who might be affected by their acts or omissions. This could include cleaners, caretakers, outside contractors and other visitors. All teachers and technicians:

- have responsibility for each other's safety, and should warn each other of hazardous situations
- are required to co-operate with the safety policies of the LEA, school, and department
- should set a good example, for example by wearing eye protection when needed
- should check that a proposed activity is in accordance with the General Risk Assessments (see section 8)
- should make sure that laboratories and prep. rooms are locked when unoccupied (unless this would block a fire exit).

The training of technicians, and especially the induction of new ones, is often ignored. Yet of the small number of serious accidents which do get reported, a disproportionate number seem to involve them. Teachers are often unaware how limited is the knowledge of some of those working as technicians in schools.

Teachers, because of their status, have additional responsibilities. Science teachers should:

- take all relevant safety precautions when preparing lessons
- check that technicians know how to carry out safely hazardous operations which they have asked them to do (eg diluting concentrated sulphuric acid)
- issue pupils with the departmental safety rules every year, and explain what they mean and why they are necessary
- remind pupils about the safety rules from time to time, and whenever a hazardous situation arises
- never leave pupils unsupervised in the laboratory (see Section 6).

6 Pupils

6.1 Rules for pupils

Pupils should be in no doubt about the standards expected of them in the science laboratory. Safety rules should be issued to them at the start of each year, and should be strictly enforced. Different rules may be needed for different age groups. (See Appendix C of reference [1]). Staff need to teach pupils actively about the rules, and the reasons for them, and should constantly reinforce them. There need to be frequent reminders, eg about wearing eye protection. The more pupils understand about the reasons behind the safety rules, the more they are likely to obey them. It may be possible to get pupils to formulate their own rules, and to negotiate a contract with the class. It is sensible for a department to develop its own style for alerting pupils to safety points on worksheets, etc, for example, by using the standard hazard warning symbols.

6.2 Know your pupils

Teachers need to be aware of the medical conditions of their pupils, especially where this might impinge on laboratory activities, eg asthma or epilepsy. Staff will need to be sensitive about the way in which they use such knowledge.

6.3 Supervision of pupils

There is little by way of national legislation governing the supervision of pupils in science laboratories. This section is based on what many teachers would consider to be good, safe practice. In some cases it is enshrined in local authority Codes of Practice.

Laboratories and preparation rooms should be kept locked at all times, unless occupied, or needed as fire exits. A court might well find that a school had some legal liability if a pupil was injured as result of experimenting with chemicals stolen from an unlocked laboratory. To some extent, there is an obligation in law to protect people from the consequences of their own stupidity.

Pupils should be allowed in a laboratory only under the supervision of a qualified science teacher, and, however senior they are, they should not be allowed to work unsupervised in laboratories. In the case of A level students, or those using a photographic darkroom, it may be satisfactory to have a science teacher within earshot. Before allowing this to happen, the teacher must be satisfied that s/he knows what the pupils are doing, that they fully understand what they are expected to do, that they are competent to do it, and are sufficiently reliable to be permitted this limited degree of supervision. It should be borne in mind that any General Risk Assessments adopted by the employer (see section 8 below) will have been drawn up on the assumption of proper supervision.

It follows from the above that laboratories should not be used as form rooms. Where shortage of space makes this an impossible ideal, the form tutor should be a science teacher, fully conversant with the rules about supervision.

6.4 Class size

The ASE has a policy [10] that classes for practical science should not exceed 20 (14 in A level groups). In Scotland, classes in practical subjects, including science, may not exceed 20. In other parts of the UK there is no limit: despite the myths, there is not one in any subject area (including art or design technology), although there may be local agreements in some places. Clearly, adequate supervision of pupils will be harder in larger classes, and this may therefore inhibit the practical activities which can take place. There may be particular issues arising from the investigative activities required by Science National Curriculum Attainment Target 1 [11].

There may be special concern where the size of the class appears to result in an overcrowded laboratory. A useful rule of thumb [12] allows 2 square metres of free floor area (ie after subtracting the area occupied by benches, cupboards, tables, etc) per 11 year old pupil, rising to 3 square metres for a 16 year old one. This gives only a rough guide, and area is not the only consideration. The accessible length of benches and tables is also relevant.

If, in the professional judgement of the science teacher, there are too many pupils to supervise safely, or the room is too crowded for safe practical activities, then the teacher should consider alternative strategies. This might include organising matters so that only a part of the class does practical work at any one time.

7 Qualified Science Teachers and Those in Training

7.1 Teaching practical science

Only qualified science teachers should teach practical science, and normally only qualified science teachers should teach in laboratories. If a non-science subject has to be taught in a laboratory, or if a non-scientist has to cover a science class in one, then:

- the Head of Science should be aware of what is going on, and be asked to accept some supervisory responsibility
- a qualified science teacher should be within earshot
- standards of behaviour should be no less than those expected in a science lesson, and science laboratory rules should apply
- no practical work should take place.

In this context, a qualified science teacher would be a person with Qualified Teacher Status, and a teacher's certificate, degree, or equivalent qualification in which a science or sciences form an appreciable part of the course, normally as a main subject.

Where an exception has to be made to this rule, either because of staff shortage, or because of a positive commitment to cross-curricular teaching, the school must make explicit arrangements for the safety training of the staff concerned. This might take the form of a regular time-tabled meeting with a senior member of the science department, with an opportunity for trying out practical activities.

7.2 Support for new teachers and those in training

Safety training, both school-based and institution-based, should form an integral part of the development of licensed, authorised and articled teachers, of students on teaching practice, and of instructors. All such staff should work under the direct supervision of a qualified science teacher. Any practical work they are proposing to carry out should be talked through in advance. Unless the qualified teacher is actually in the room, they may be prohibited from carrying out certain types of activity (eg with radioactive sources).

Newly qualified teachers also need considerable support in their first year. One or more members of the departmental staff should be designated to meet with them on a regular weekly basis, to talk through their proposed practical activities, and induct them into the safety procedures of the department.

Experienced teachers, new to a particular school, may also need a period of induction. Different employers do have different local rules, and it will be necessary to explain to new staff what this department's procedures and policies are.

8 The COSHH Regulations

8.1 Scope of the regulations

The *Control of Substances Hazardous to Health (COSHH) Regulations 1989* are intended to protect employees and others from substances which might be hazardous to their health. As far as schools are concerned, this includes micro-organisms and all uses of chemicals classified as harmful, toxic, very toxic, irritant or corrosive. The chemicals might be intermediates in or products of a chemical reaction. Dusts at substantial concentrations, and other substances with 'Maximum Exposure Limits' or 'Occupational Exposure Standards' are also covered, but science teachers are unlikely to meet any. Strictly speaking, explosive, flammable or radioactive substances and oxidising agents are not covered by the COSHH Regulations. However, flammable substances cause more problems in school science than most other categories, and the distinction between COSHH and non-COSHH chemicals will probably be lost on most teachers. Therefore the ASE Laboratory Safeguards Sub-committee recommends that teachers treat all chemicals as if they were covered by the Regulations.

8.2 Risk assessments

Under the COSHH Regulations, employers are obliged to carry out a Risk Assessment before hazardous substances are used or made. Because of the enormous variety of procedures used in education, it is not practicable for employers to carry out a formal risk assessment for every single activity. Therefore, the Health and Safety Commission [7] has endorsed the use of *General Risk Assessments*. These could be written by the employers themselves, but the great majority have followed HSC advice and adopted various nationally available publications such as references [2, 3, 9].

It is important to draw a distinction between hazard, and risk. The hazard is inherent in the nature of the material. The fact that the substance is hazardous alerts the teacher to the possibility that there may be a risk from using it. The extent of that risk depends on the exact circumstances for the proposed activities.

Where an employer has adopted *General Risk Assessments,* the role of the teacher is to compare their proposed procedure with that defined in the *General Risk Assessments.*

- Is this activity educationally necessary?
- Is there an alternative, less hazardous, substance or procedure?
- Is it a teacher demonstration, or pupil practical work?
- What is the age and experience of the pupils?
- How will pupils be warned about the hazards?
- What is the scale of working?

- What personal protection or control measures are necessary (eye protection, gloves, fume cupboards, safety screens)?
- How will residues be disposed of at the end, without risk to the technicians?

Risks may be reduced by making sure that teacher, pupil and technician understand what those risks are.

There is no requirement for schools to produce their own Risk Assessments, or to write them down, as long as there are written *General Risk Assessments* available, although safety officers in some LEAs may try to argue otherwise. Unfortunately, HM Factory Inspectors have sometimes found that schools claim to be using, say, *Hazcards* as the basis for their *General Risk Assessments,* when that is demonstrably not the case. For example, they may be using chemicals considered unsuitable for pupils of that age, or they may be working on a larger scale than suggested.

Schools need to be able to demonstrate that they are acting upon the *General Risk Assessments* their employer has adopted. Some schools laminate their CLEAPSS *Hazcards,* and give technicians a standing instruction to issue them with trays of equipment and chemicals whenever relevant: some have a rubber stamp made, saying "Check Hazcard", which they then use to draw attention to problems throughout teachers' and technicians' guides. However it is done, the ASE Laboratory Safeguards Sub-committee strongly urges schools to annotate their syllabuses, schemes of work, work-sheets, courses guides, teachers' and technicians' notes, etc to point out hazards, necessary precautions, and any other safety requirements. Many published schemes will already have done part of this job for the school, but many were written before the legislation was implemented or its consequences understood, so the school must check that the proposed procedure does comply with their current *General Risk Assessments*. The school should then indicate by suitably annotating the text that this check has been carried out.

8.2.1 Special risk assessments

If situations are not covered by the published *General Risk Assessments,* perhaps because unusual chemicals are being used, it may be possible to derive a Risk Assessment from those available for comparable materials. In general, however, a *Special Risk Assessment* will be necessary, and the employer should have defined a procedure for obtaining this, perhaps by reference to a local committee of suitably experienced and qualified teachers and/or occupational hygienists, but in many cases, for members, it is likely to involve contacting the CLEAPSS School Science Service. A *Special Risk Assessment* may also be necessary if, for example, a school does not have an efficient fume cupboard, and wants to know if some procedures can be carried out in the open laboratory: the answer may well be yes, on a sufficiently small scale. Teachers who are interested in the process of a *Special Risk Assessment* should consult the SSERC document, reference [6], which may be of particular value where there is extensive project work.

8.2.2 Observing risk assessments

All staff, both teachers and technicians, must clearly understand the nature of the Risk Assessments, and their personal responsibility for implementing them. If the Risk Assessment requires eye protection to be worn, the teacher must do so, and must insist that the pupils do so also. In extreme cases, failure to comply with an employer's safety rules could result in dismissal, although the teacher could normally expect to receive warning before such drastic steps were taken. Similarly, if the Risk Assessment requires a fume cupboard to be used, then it is no excuse to argue that one was not available. The procedure must not be carried out, unless a fume cupboard is used, or unless a *Special Risk Assessment* has been carried out.

8.2.3 Risk assessments beyond COSHH

Because the Health and Safety Executive considers the COSHH approach to have been a success, they are seeking to extend Risk Assessment to situations other than those involving micro-organisms and hazardous chemicals. As indicated above, the ASE Laboratory Safeguards Sub-committee believe the extension to flammable substances and other dangerous chemicals to be sensible. There are some other practical activities, for example the use of high voltages (see section 9), where Risk Assessment would also be advantageous.

Regulations are likely to be introduced in 1993, as part of the harmonisation of legislation across the European Community. This will probably require employers to carry out a Risk Assessment before any practical activity takes place. If, as seems likely, education employers adopt the same approach as with COSHH, then existing publications, such as [1, 2, 3, 4, 5] will form the basis for *General Risk Assessments* in the new areas.

9 The Danger Areas

9.1 Sources of advice

It is not appropriate, in a handbook such as this, to list every hazardous chemical, or every potentially dangerous situation. The reader is referred to more specialist publications [1, 2, 3, 4, 5, 9]. However, a small number of practical activities have resulted in a disproportionate number of serious accidents in recent years, and it is important that all science teachers should be aware of them. The ASE Laboratory Safeguards Sub-committee is always interested to hear from members about their accidents and near-misses, because only in this way can we get a picture of what is happening in schools.

The main danger areas can be categorised as:

- burns from alcohol (ethanol) fires
- electric shock during teacher demonstrations
- effects of sniffing chlorine
- hydrogen explosions
- alkali metal explosions

In a number of these cases pupils were injured, although the activity was a teacher demonstration. Often, such problems arise because pupils are sitting too close to the demonstration bench, and are too close together to jump back in the event of an accident. Leave a gap of 2 – 3 m between the bench and the pupils.

9.2 Ethanol fires

In the mid-1980's, half a dozen pupils were badly burned (to the extent that they needed skin grafts), in separate incidents involving alcohol (ethanol) fires. Most involved model steam engines, being demonstrated by the teacher. The details vary, but typically the teacher thought that the flame had been extinguished, because the fuel had been used up, and attempted to refill the burner, only to find it was still alight. Other incidents have arisen when using ethanol to extract the chlorophyll from leaves, generally because the teacher has not appreciated just how flammable ethanol is.

Whatever the cause, the message is very clear – ethanol is very dangerous. It should *never* be heated with a naked flame: instead, use a bath of water, which has itself come from a hot tap, or been heated in an electric kettle. *never* use ethanol as a fuel in model steam engines.

9.3 The high voltage transmission line

In recent years about half a dozen teachers have been the victims of electric shock. These incidents all arose from a particular demonstration in which a 12 V power source is stepped up to about 240 V, transmitted for some distance down a pair of wires, and then stepped down again. The purpose is to demonstrate why power is transmitted through the national grid at a high voltage. In each incident, the teacher, intending to make some adjustment, grabbed hold of conductors (usually crocodile clips) live at a potential of about 240 V, and was unable to let go.

The Health & Safety Executive is particularly concerned about these electrical accidents, since after the first two or three, they issued guidance to LEAs about measures to avoid them. Essentially, they suggested that exposed conductors should not be live at a potential of more than 50 V, and that where equipment was live at higher voltages, it should be contained within a clear polycarbonate enclosure which prevented accidental contact with dangerous voltages. Details will be found in Chapter 3 of reference [2], including some suggestions for constructing safe equipment. Under RIDDOR *(Reporting of Injuries, Diseases, and Dangerous Occurrences Regulations 1985)*, all accidents which give rise to electrical burns must be reported to the HSE within seven days, and failure to do so in some of these incidents has heightened HSE concerns.

9.4 Chlorine sniffing

There have been a number of cases where pupils have suffered from the effects of chlorine. They usually resulted from the teacher asking pupils to smell chlorine which had been generated in the laboratory, without first checking whether any pupils were asthmatic or showing them how gases may be safely sniffed. A small number of pupils seem to be particularly sensitive to tiny amounts of chlorine.

To avoid problems, teachers should check that pupils are not asthmatic, and must in any case warn pupils not to breathe deeply. Pupils (and their teachers) need to be shown how to smell gases safely. This involves first filling the lungs with air (so that a deep breath cannot be taken), pointing the test tube away from the face at a distance of 10 – 15 cm, then using the other hand to gently waft vapours to the nose. If pupils cannot be trusted to do this, they should not be allowed to carry out the activity. Techniques for smelling gases (and for other potentially hazardous activities) should be taught in low hazard situations, before the need to use them arises.

9.5 Hydrogen explosions

Mixtures of air and hydrogen are explosive over a wide range of compositions, and can be ignited not only by naked flames, but also by catalysts, such as transition metals and their oxides, at temperatures below red heat. The one prosecution of a science teacher already mentioned (p 131) was the result of a hydrogen explosion. In attempting to reduce metal oxides by hydrogen, it is far safer to take hydrogen from a cylinder of compressed gas than to generate it, because the faster flow flushes air out more quickly. Both teacher and pupils should wear eye protection, and should be protected by safety screens. Before igniting the hydrogen, or heating the metal oxide, a sample of the issuing gas should be collected in an inverted test tube, and ignited some distance away. If it is safe to ignite the issuing gas, the gas in the test tube will burn for a sufficient length of time to act as a torch.

9.6 Alkali metals

The vigour of the reaction between alkali metals (especially sodium and potassium) and water sometimes surprises teachers: pieces of hot corrosive metal may fly out of the vessel. Pupils should not sit closer than 2 – 3 m, and both teacher and pupils should wear eye protection and be protected by safety screens. Only small, rice-grain sized pieces should be used.

10 Open-ended Projects and Investigations

National Curriculum Attainment Target 1

The safety implications of pupils having to design their own investigations are considerable. Although they can be encouraged to consider safety as part of their planning, and given guidance on how to do this, the teacher must obviously vet the plans before they are implemented; not an easy task, with limited time available, a class of thirty or more pupils, and the implications of a pupil's plan not necessarily very clear – to the teacher or the pupil.

Peter Borrows is General Adviser (Science) for the London Borough of Waltham Forest, before which he was a science teacher for over 20 years.

References

[1] ASE (1988) *Safeguards in the school laboratory,* 9th edition, ASE.

[2] ASE (1988) *Topics in safety,* 2nd edition, ASE.

[3] CLEAPSS (1989) *Hazcards,* CLEAPSS.

[4] CLEAPSS (1988) *Laboratory handbook,* (and later supplements), CLEAPSS.

[5] SSERC (1979) *Hazardous chemicals: A manual for schools and colleges,* Oliver & Boyd.

[6] SSERC (1991) *Preparing COSHH risk assessments for project work in schools,* SSERC.

[7] HSC (1989) *COSHH: Guidance for schools,* HMSO.

[8] Vincent, R. & Borrows, P. (1992) *Science department safety policies,* School Science Review, vol 73, no 264.

[9] HMI (1990) *Microbiology. An HMI guide for schools and further education,* 2nd impression, HMSO.

[10] ASE (1992) *ASE policy, present and future,* ASE.

[11] DES (1991) *Science in the National Curriculum,* HMSO.

[12] ASE (1989) *Building for science: A laboratory design guide,* ASE.

Useful addresses

CLEAPSS School Science Service, Brunel University, Uxbridge, UB8 3PH Tel 0895 251496

(CLEAPSS = Consortium of Local Education Authorities for the Provision of Science Services: was CLEAPSE)

SSERC 24 Bernard Terrace, Edinburgh, EH8 9NX Tel 031 668 4421 (SSERC = Scottish Schools Equipment Research Centre: was SSSERC)

Note: the services of CLEAPSS and SSERC are only available to teachers whose education authorities or schools are members.

8 The Use of New Technology in Science Teaching

Maurice Tebbutt

It has become conventional to use the term 'New Technology' to mean the rapidly burgeoning array of equipment which uses micro-electronics – electronics which is based on the use of highly miniaturised and very powerful circuits encapsulated in 'chips'. This has been evident in the domestic market, with the appearance of ever smaller, ever more versatile hi-fi equipment, televisions and video recorders. It would be surprising if this had not rubbed off directly on to science teaching.

'New Technology' has had at least the same sort of impact in the commercial market, generating enormous changes in reprographics, communications and, particularly, in the use of computers. In less than two decades, computers have progressed from being solely large, expensive machines, operated by experts on behalf of the 'user', to become predominantly small, powerful, fast, versatile, relatively cheap, owner-operated, user-friendly ones. As a result of the huge increase in the numbers of computers and users, software has been written to cater for a whole range of uses which were not even considered only 20 years ago. Thus, while personal computers can 'crunch numbers' as in the past, they can now also process and manipulate text, data, and graphics; and communicate with other computers or with apparatus of various kinds. Hence one problem of writing about this area is that the technology makes it possible to do things which have not hitherto been considered possible: another is that it seems to generate in users new ideas about its capabilities, and at a pace which makes it difficult to predict what will be available in 5 years, let alone another decade.

For these reasons this chapter has been divided into three sections, established areas, less established areas with potential, and practicalities.

In writing on this topic, it is necessary to consider a variety of hardware, partly because there are a number of machine manufacturers, and partly because machines have been introduced over the years and continue to be used for different times after their introduction. Figure 1 shows the computers which have been used in schools during the last decade or so in this country.

The education market has been dominated by two firms which are relatively unknown in the commercial sector; Acorn and Research Machines. The implementation of computing in schools was originally aided by the interest of the

Time	Acorn	Research machines	IBM	Apple
	BBC 'A'	360Z		
	BBC 'B'			
	BBC 'Compact'	460Z		
	BBC 'Master'		PC	
	Archimedes	Nimbus		Macintosh
	A300 series	186		
	A400 series	286		
	A3000 = 'New BBC'			
	A500 series	386		
	A5000			
▼	▼ ▼	▼	▼	▼

Figure 1 How computers have been used in schools over the 10 years 1982–92

British Broadcasting Corporation, who were closely involved in the development of the Acorn machine, which subsequently took the name 'BBC', and who used it in a series of television broadcasts on computing. As is common with all computers, this machine has had a number of different versions, which are more or less compatible with each other. At the time of writing, there are still very many BBC B and Master machines in use in schools. While they are still giving good service, it is certain that over the next year or two they will need to be replaced. The direct replacement would be the 'new' BBC machine, the Acorn A3000, but there are likely also to be the other versions of the Archimedes machine (see Fig 1) in schools. These have differed in speed and memory capacity, but are generally compatible in terms of software.

Probably the next most popular machine in schools today is the Research Machines Nimbus. These have hitherto been supplied in three forms called the 186, 286 and 386 – the labels representing the main processor chip used in the machine. Since these machines have been increasingly compatible with IBM PC machines, they have access to the wide range of software developed worldwide for these. This is also true of other PC compatible machines, manufactured by IBM or other firms, which have been adopted by some schools. Others have adopted the Apple Macintosh range: these sometimes use their own, distinctive, software, but they can also use some of the same software as IBM PC machines.

A problem which follows from the range of hardware which is used in schools is how to describe the software which is available. I have tried to be sufficiently specific to be helpful to the reader without overburdening the text. In sections 1 and 2 examples are given for the BBC/Archimedes range, while other examples, together with equivalents for other machines are given in section 3.

1 Established Areas

1.1 Videos

Video tape recorders are so well established in the domestic arena that nothing needs to be said about the hardware, and its advantages, but it is worth making some points about software. The format of videos is not amenable to change, and they tend to be both too long and written to match a particular programme format. These disadvantages, and the problems of moving video machines and monitors, make it difficult to integrate them into lessons, and this is liable to encourage teachers to use them more as 'time filling' activities. It would be better if one could show sections of videos, on machines permanently installed in laboratories or mounted on trolleys. Some material of this type, with short sections which are relatively 'free standing' and address one of two concepts per section, is currently in preparation by Longman.

1.2 Simulations

Some aspects of science and technology cannot conveniently be included in courses because the object or system which is of interest is too large or too small to be considered in reality, too dangerous, too complex, or too inaccessible or impractical. A simulation may be the answer. They may also be used to model school experiments, or to establish abstract concepts.

Computers are ideally suited to producing models of reality which can be used to investigate the behaviour of the real thing, and many have been produced for schools. Simulations written for comparatively low-powered computers like the original BBC machine, have been relatively more popular than those produced for more powerful machines like the Nimbus and the Archimedes. It remains to be seen whether software publishers are prepared to expend the greater effort needed to program these machines in order to produce simulations, or whether they will prefer to devote themselves to writing more general-purpose packages. Those which have been produced, such as *Orrery* for the Archimedes, are good, if expensive.

Examples of the topic areas covered by the simulations which are available are shown in figure 2.

1.3 Datalogging

1.3.1 Datalogging itself

How many science teachers have wished that they could easily perform those experiments, notably in Biology, which last much longer than the school day, and which may involve measurements of a number of parameters? How good it would be to be able to measure the temperature, *pH* and oxygen concentration in a pond or tank over a number of days, and show how these are related to the light intensity. Fortunately this is now possible, by means of a technique called datalogging. *Datalogging* is the collection and recording of data which is varying with time, and it is increasingly being seen as one of the most effective uses of computers within science education. It also makes possible experiments with a shorter time scale than could normally be investigated, for example what happens when you drop a magnet through a coil. Figure 3 shows a selection of possible experiments and the parts of the National Curriculum to which they contribute.

More means for performing datalogging have become available as it has become more popular, and this has led to it becoming more complex. As will become apparent, this complexity is reflected in the increased range of the elements of datalogging which are now available – computers, dataloggers, software, and transducers. The principles remain the same, however, and are shown in figure 4.

Datalogging can, in principle, be used for any experiment which would currently be used in a school science course and introduces extra possibilities. Variations in parameters such as light intensity, temperature, pH, or oxygen concentration can be monitored. The next stage is to convert these variations to variations in an electrical parameter to which the third stage of the process, a datalogger, a computer or a combination of the two, is sensitive. In general, this means that the transducer needs to produce a voltage, either directly, or by means of some processing of the signal.

Topic Areas	Simulations
Systems which are too large to study in reality	Power stations B, A Electrical power consumption of a town or village B, A Solar system (various) B, A Earth's Biosystem (SimEarth or Ecos) PC, A
Systems which are too small to study in reality	Kinetic theory eg. Molecules in motion B Particles and diffusion B, N, PC
Systems which are too complex to study in reality	Water Pollution B, N Fuel consumption of a car in motion B Manufacture of sulphuric acid; Siting an aluminium plant B Population Genetics B, A, N, PC Process simulation A Domestic Heating
Systems which are too dangerous to study in reality	Nuclear power station B, A, N, PC First Aid B, A, N Investigating Radioactive Sources B, A, N
Systems which are inaccessible or impractical	Motion in space B Action of the eye B, A, N Menstruation and pregnancy B Human blood groups B Insulin B
Experiments which can be performed in schools	Titration B Worcester circuit boards B, N, PC Spirometer B, A Sampling in Ecology B, N, A, PC Reaction Rates B, A, N
Simulations which establish abstract concepts	Waves B, A Lake Web B Particles and diffusion B, N Fourier Analysis B, N Decay B

Key: *B = BBC B or Master; A = Archimedes N = Nimbus; PC = IBM PC or compatibles*

Figure 2 Topic areas covered by simulations

Experiment	National Curriculum SoA/PoS	Transducer(s)/software
1. Investigating photosynthesis in a pond plant	NAT2.7b NPS2.III/IV	Oxygen sensor, possibly light sensors; powerful software
2. Investigating respiration rate of small animals	NAT2.3a NPS2.III	Oxygen sensor or electronic manometer; simple or powerful software
3. Detecting energy release during germination, respiration or combustion.	NAT2.3a NPS2.III	Temperature sensor; simple or powerful datalogging software, or 'multimeter' program
4. Enzyme catalysis	NPS2.III, 3.III	Colorimeter; or commercial light sensor; or home made colorimeter; or pH meter or sensor – all depending on the reaction; powerful software
5. Investigating plant growth	NAT2.3a, 2.3d, 2.7b; NPS2.III/IV	Movement sensor; possibly light, temperature sensors; simple or powerful software
6. Investigating osmosis	NAT2.9b; NPS2.IV	Electronic manometer
7. Modelling animal huddling		Temperature sensor; powerful software
8. Investigating reaction rates	NAT3.7d; NPS3.III	Home made or commercial colorimeter; or movement sensor coupled to gas syringe (depending on experiment); simple or powerful software
9. Plotting pH curves in real time, or thermometric titrations	NAT3.5c, 3.6f; NPS3.III/IV	pH meter or sensor, temperature sensor; simple or powerful software
10. Examining the combustion process	NAT3.5d, 3.6d; NPS3.III	Oxygen sensor; powerful software
11. Investigating oscillations and damping	NPS4.IV	Movement sensor; simple or powerful software
12. Investigating cooling, and the efficiency of insulators; examining cooling curves	NAT3.7f, 4.7d NPS3.III, 4.IV	Temperature sensors; simple or powerful software
13. Investigating the absorption of thermal radiation	NAT4.7c; NPS4.III/IV	Temperature sensors; simple or powerful software
14. Examining the decay of thoron or Pa^{234}	NAT3.9f; NPS3.IV	Electrometer or ratemeter sensor; simple or powerful software

Figure 3 Experiments using computers

In practice, the process is a little more complicated than that shown in figure 4. The difference between figure 5 and figure 4 is in the detail of the logging system. The logging can be done by some computers, provided that they are equipped with the appropriate circuitry, whose function is to convert the analogue voltage produced by the transducer (which varies continuously) into a digital signal (chopped into bits) which the computer can process. (BBC machines, unusually, incorporate this device, but it is an extra for most machines)

However, in each case it is necessary to provide a 'buffer box' to protect the computer in case inappropriate voltages are applied to it. These can be bought from specialist or laboratory suppliers, or made by hand, since it is quite easy for someone with electronic construction skills. Alternatively, the logging system can consist of a Data Logger which can operate on its own most of the time, but can be connected to a computer

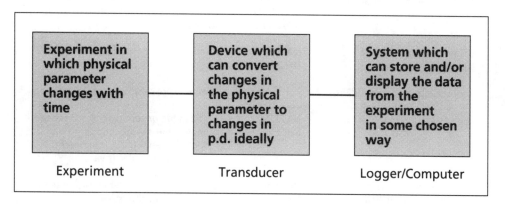

Figure 4 The principle of datalogging: simplified

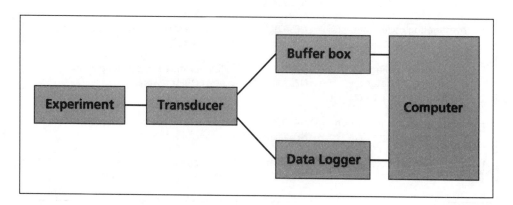

Figure 5 The principle of datalogging

when it is necessary to process the data that has been collected. Since Data Loggers incorporate protection circuitry as well as the circuitry needed to convert analogue to digital signals, they can also be used when directly connected to computers even if the computers do not themselves have such protection.

Although the principle of transducers is simple, they convert a physical parameter into an electrical one, the range of possible transducers adds to the apparent complexity. Figure 6 shows how they can be grouped.

Groups 1 and 2 are useful for those wishing to start datalogging, since the equipment may already be available in schools, whereas Group 3 requires expenditure which is often considerable. In contrast, group 4 can lead to substantial savings, but it requires both knowledge and some practical skills.

1.3.2 Software for datalogging

External software is needed for both data gathering and processing with most computers and dataloggers; a few dataloggers incorporate software internally for doing the data gathering only. Figure 7 summarises the characteristics, advantages and disadvantages of the different categories of software which are available.

Types of Transducer

Commercial apparatus

Group 1   **Standard apparatus,** such as pH meters, colorimeters, electrometers

Group 2   Sensors from **environmental monitoring systems**

Group 3   **Dedicated apparatus.** Sensors intended for particular dataloggers, or sensors which can be used with any datalogger or computer

Non-commercial apparatus

Group 4   'Home made' transducers

Figure 6 Types of transducer

	Simple	**Powerful**
General purpose	**Char.** Basic program can be used with a wide range of sensors in order to conduct a wide range of experiments. The program allows variation of a limited number of parameters, such as numbers of channels, recording time, sensitivity, but without further data manipulation facilities. **Example.** SCOPE (Tebbutt (1991)) **Adv.** *Easy to use. Less chance of infrequent users forgetting what to do. Useful stepping stone to more powerful software.* **Disadv.** *Limited range of capability, eg. unable to plot traces in different colours; or print screen.*	**Char.** Program with many facilities to aid collection and processing of data. Can be used with a wide range of sensors to perform many different experiments. Usually controlled through a succession of menus. **Example.** DATADISC+ (Philip Harris), INSIGHT (Longman-Logotron) **Adv.** *User has available a vast range of resources for collection of data, such as calibration of inputs; and data processing, such as replotting, at will, reading data points easily from the trace, or plotting mathematical functions of the data.* **Disadv.** *Can be daunting to learn to use, and easy to forget for infrequent users. Danger of not seeing the wood for the trees.*
Dedicated	**Char.** Software which is intended to be used with a specially made piece of apparatus designed for an individual experiment. **Example.** "Leicester" modules (Deltronics) **Adv.** *The software is tailored to the hardware, so the whole system is easy to use. The system is likely to be usable even by teachers or pupils who are inexpert* **Disadv.** *The system lacks flexibility – only the components supplied can be used. If components fail, exact replacements are needed. The system can be quite expensive on a cost/experiment basis*	

Figure 7 Advantages and disadvantages of datalogging software

1.4 Spreadsheets

What can you do with spreadsheets in science? Figures 8, 9 and 10 show examples of worksheets which aim to use spreadsheets.

The most obvious use is in helping to gather, process and display the data from experiments (figure 8). This is particularly useful for experiments where:

- a good deal of processing is required,
- the pupils do not need to be able to do the processing
- groups of pupils can contribute to the class results, which all go into the spreadsheet

They may also simply help in organising the data or allowing pupils to perform quite elementary processing, perhaps as part of an investigation. Most spreadsheet packages allow the results of the processing to be displayed in graphical form, but the range of options available for this is sometimes rather limited, betraying perhaps the business pedigree of most packages.

Clearly, spreadsheets are also well suited to the manipulation of data, and such activities also find a place in Science Education (Figure 9).

Examples are:

- the calculation of electricity, gas or telephone bills, given data on meter readings
- the calculation of the quantity of electricity or water used by a household over a period of a day, week, month or a year, based on observations made by pupils over a relatively short period;
- calculation of the energy loss from a house under certain heating conditions;
- the comparison of the proportions of various forms of refuse produced by households today and at various times in the past, based on data, or measurement.

Any of these results could be graphed.

The final group of activities is best described as modelling tasks. These might involve mathematical modelling of the discharge of a capacitor or radioactive decay; or determining the value of the load resistor needed to ensure the maximum transfer of power to a load from a source. It is also possible to use spreadsheets to explore the design of animals and leaves (Figure 10).

Spreadsheets consist of an array of cells arranged, predictably, in rows and columns. The cells can contain alphabetic data, such as labels; or numeric

Reaction timing

You can see how long your nervous system takes to react to an event (this is called the reaction time) using just a ruler, as shown in the diagram.

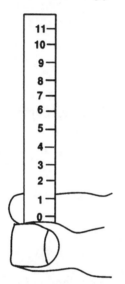

Someone else holds the ruler up and you put your thumb and finger at each side of the bottom of the ruler, where the zero mark also is. When the other person drops the ruler, you have to catch it. See what the reading is on the centimetre scale. The computer will calculate your reaction time from the reading.

Use the spreadsheet REACTN2 to enter your results. Try to get as many people's results as possible when they are using their 'normal' hand'.

REACTN2 should look like this.

	'Normal' hand		'Other' hand	
Name	Distance	Reaction time	Distance	Reaction time

Use the data to plot a bar chart to show which is the most common value of reaction time.

Use the second set of columns to investigate people's reaction time when they are using their 'other hand'.

Are the reaction times for each hand similar or different?

Figure 8 Use of spreadsheets: Reaction Timing

The Solar System

The Solar System consists of the Sun and 9 planets. Load the speadsheet **SOLAR 1** which should look like this.

Name	radius/Mm	orbit radius/Gm	Mass in M(E)	surface temp/°C
Sun	695.1	–	332946	6000
Mercury	2.42	58	0.06	350
Venus	6.05	108	0.82	465
Earth	6.37	150	1.0	15
Mars	1.73	228	0.12	-23
Jupiter	71.3	778	317.9	-150
Saturn	59.9	1427	95.2	-180
Uranus	25.4	2870	14.5	-210
Neptune	24.27	4500	17.2	-220
Pluto	1.47	5900	0.002	-220

M(E) = Mass of Earth = 6.0×10^{24} kg

1. Use the graphics capability of the spreadsheet to plot the following:

   (a) a bar chart of the distances of the planets from the sun
   (b) a bar chart of the sizes of the planets
   (c) pie charts of the masses of the whole solar system, or just the planets

   Write a short account about what each graph tells you about the solar system

2. Plot a graph of surface temperature against distance from the Sun. Describe in words how temperature changes with distance

Figure 9 Use of spreadsheets: What is the solar system like?

<u>Level 2</u>

Designing leaves

Leaves respire through their surfaces, so that leaves are likely to develop in such a way that they have the maximum surface area for a given volume.

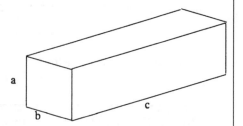

A rectangular prism is useful as a crude model of a leaf, since it is easy to calculate its volume and surface area.

Some background information

The volume of the box is a x b x c, and its surface area is 2(a x b) + 2(b x c) + 2(a x c).

Your task is to use a spreadsheet to explore the changes in the ratio of surface area/volume for prisms, or leaves, of different shapes; to see which of these shapes produce(s) a maximum value of this ratio, and to see whether these particular shapes correspond to real leaf shapes

Load the spreadsheet "LEAF2". It should look like this:

	A	B	C	D	E	F
00 :	side a	side b	side c	volume,V	surface area,A	A/V
01 :	10	10	10	1000	600	0.6
02 :						

Alter a,b or c and see what effect this has on A/V. The changes you make will usually alter V too, so you should try to keep V approximately constant like this:

	A	B	C	D	E	F
00 :	side a	side b	side c	volume,V	surface area,A	A/V
01 :	1	31.6	31.6	1000	2126	2.13

What shape(s) has the largest ratio A/V? Try to make this ratio as big as possible.

Do these shapes resemble real leaves at all?

Figure 10 Use of spreadsheets: Designing leaves

data, that is numbers; but the most important feature is that the cells can contain formulae, which enable data from other cells, or constant numbers, to be combined mathematically. If subsequent changes are made to the data, the spreadsheet automatically recalculates all the cells. Spreadsheets were originally designed for use in business, and although at first sight they do not have obvious applications in Science Education, in practice there are many suitable applications.

It will be apparent that these examples are taken from different levels of work: some are appropriate post-16; others to years 7 or 8. However, at any particular level, the difficulty of an activity can be varied by adjusting the amount of help which is given.

Lowest

1. Spreadsheet framework and data (and possibly formulae) are in the machine. Pupils merely use it by searching for patterns or plotting graphs.

2. Spreadsheet framework (and possibly formulae) in the machine. Pupils enter data and use the spreadsheet by searching for patterns or plotting graphs.

3. Spreadsheet framework and data are in the machine. Pupils enter formula(e) and use the spreadsheet by searching for patterns or plotting graphs.

4. Spreadsheet framework is in the machine. Pupils enter data and formula(e) and use the spreadsheet with guidance.

5. Pupils design the framework, enter data and formula(e) and use the spreadsheet with guidance.

6. Pupils design the framework, enter data and formula(e), and use the spreadsheet entirely on their own.

Highest

Figure 11 Use of spreadsheets: varying the level of difficulty

Figure 11 shows how this might be done by providing the pupils with a spreadsheet which is completed to different extents. At the lowest level of difficulty the pupils are expected only to use the spreadsheet which they are given, but they are required to do more and more as the difficulty level rises, until, at the highest level they are expected to design and operate the spreadsheet entirely without help.

One advantage of spreadsheets is that the software is rather less complex than for some other aspects of IT in Science. Almost any spreadsheet package to which a Science Department has access will work well, but it will be necessary to consider user friendliness, particularly if the package is to be used by younger pupils, and the graphics capability. *Grasshopper,* which was originally intended for primary schools, is sufficiently simple and therefore friendly for many pupils, and there are versions available for the original BBC machines, the Nimbus and the Archimedes. The packages written for the latter machines are much more powerful, but may be difficult for younger pupils, or for beginners. Examples are *Excel* for the Nimbus (and other PC compatibles, as well as Apple Mac) and *Schema* or *Eureka* for the Archimedes. Alternatively, *Datasweet* and *Datasheet* are packages written for the Archimedes and intended for primary schools.

1.5 Databases

Databases are like filing systems which are held in a computer instead of on cards, and the terminology which is used is similar. Thus, a database on school science apparatus would consist of a collection of records which are equivalent to cards. Each record contains sections, called fields, containing data on a number of aspects of the apparatus: (eg whether it is mains powered or battery powered; if it is glassware; whether it is general purpose or specialised; its size or weight; or its general function.)

Using databases it is possible to answer questions posed either singly, or in combination. Thus, in this example, it would be possible, by asking one suitable question, to identify apparatus which could fulfil a particular function; or, by asking a combination of questions, find electrical apparatus which was portable, by virtue of being powered by batteries and weighing less than some specified amount. Alternatively, it would be possible to see how the masses of the apparatus are distributed by plotting a pie chart showing the relative masses of the apparatus.

Database packages consist of two parts – a database management program which controls the interrogation and presentation of the data; and one or more datafiles, which contain the data which is to be interrogated. The database management program is a general purpose one, which is not specific to any subject, whereas the datafiles do cover specific topics. A popular database is called Key, (or Keyplus in its more recent form), and in one of its forms it can be used with the BBC machine, the Nimbus, the IBM PC, and the Archimedes. It has a wide range of datafiles, and those in Science include:

Our neighbours in space
Energy
Rocks and minerals
Mammals of the World
Minibeasts of Britain and Ireland
Periodic table

The most straightforward method of using this, and other databases, is to interrogate one of the pre-existing datafiles, such as those listed above. This enables the pupils to have practice in deciding what they would like to know about the subject of the datafile, and to develop the skill of asking appropriate questions. Although these questions are not of the same kind as those which need to be posed in the initial stages of an investigation, the process of questioning is clearly similar. Since pupils find it difficult to ask questions in a testable form, it is helpful to have a situation in which they can practise.

An alternative mode of use for databases is for pupils to gather the data for the datafile before subjecting it to analysis. The pupils therefore have to have a good idea of the kinds of question which they might ask even before the database exists. This is much more demanding than the first mode of use, because they have to decide which fields would be appropriate, and they must specify these with sufficient precision to avoid overlap between fields which would lead to ambiguity and confusion in subsequent use of the database. Most database management programs incorporate the ability to construct a datafile, as well as that of interrogating one.

Increasingly school science courses are being linked with the world outside school. One way of doing this is to make use of large national databases, such as the *National Environmental Database*. This is done by connecting the school computer with the database computer via the public telephone service, which requires an extra piece of hardware, called a *modem,* and some communications software to control the process. Another way is to use software such as *Pinpoint* (for the Archimedes) which combines the collection of data by questionnaire, its processing and its presentation.

1.6 Reprographics

It is not proposed to go into great detail here about the use of the photocopier by Science departments! However, it should not be forgotten what a liberating influence it has been to have the facility to reproduce materials quickly and with high quality. Increasingly, it will be possible to do so in colour. The quality of the original now matters much more than it once did – the quickly scrawled page of instructions will not bear comparison with material which is produced commercially, and pupils tend to be more influenced by the apparent lack of quality of the hastily produced worksheet than they are by the good quality of the alternatives. (In other words, once again teachers have to run harder even to stay level!) These developments have placed pressure on teachers to produce high quality originals. Fortunately, publishers are becoming aware of the problems of hard-pressed teachers and more materials are appearing which incorporate in their cost an explicit or implicit licence to copy (usually within the purchasing institution). An alternative to these methods is to produce material 'in house', but the pressure on quality remains, and can probably only be satisfied by using Desk Top Publishing, which is described in the next section of this chapter.

1.7 Wordprocessing/DTP

Wordprocessing is one of the major applications of computers in the world at large. It is characterised by the ability to type material on a keyboard like that of a typewriter, but instead of the typed material being produced immediately, it is shown on a screen, only later to be printed on paper, or stored in some kind of memory. The material on the screen can be added to, deleted, corrected, or sections of text can be moved about from place to place. This *editing process,* allows written material to be developed without the need to start again and rewrite everything after the initial attempts have been examined, thought about and perhaps discussed. The production of written material can therefore proceed by the process of draft and re-draft. Of course, it is not likely to be worthwhile for science teachers to take the time and responsibility for teaching pupils how to use wordprocessors. This needs to be part of central school policy, which is supported by a range of departments, who could each set appropriate activities designed to reinforce and use the pupils' basic skills: draft and re-draft, is, for example, emphasised in the English National Curriculum. Then wordprocessors could ensure that the writing done in science lessons is much less routine than is usually the case. Pupils could occasionally do more extended writing, and they may be more motivated to do so since redrafting just means editing. Moreover, the finished product can be of high quality, since many wordprocessor packages incorporate spelling checkers, and all allow the final printing to be produced from the final version of the written material, stored on disc.

In science, pupils might therefore work in groups on different aspects of topics like the Earth in Space, the electromagnetic spectrum, or the uses of radioactivity, researching and writing up sections of the work for duplication and distribution, or display, to the rest of the class. The teacher's role is to encourage the pupils to research effectively, and to write in such a way as to communicate with their peers.

As might be expected, there are many wordprocessor packages available for the commonly used machines. The most familiar one for BBC machines is probably *View,* since it was supplied as standard with many of the most recent Master machines. PC compatible machines like the Nimbus have a wealth of choice, but the most popular packages seem to be *Word,* either as a free standing program, or as part of the *Works* package, or running within the *Windows* environment. A common package for the Archimedes is *1st Word Plus* which has been supplied as part of the *Learning Curve* (a package of hardware and software aimed at parents and schools). All of these programs make it easy to see how the text will look when it is printed since they show the text on the screen in this way.

Desk Top Publishing (DTP) is at first sight merely an extension of word-processing, but, in practice, it provides such an increase in facilities that it should be regarded as a separate feature of computers. The demands of these packages on processing and memory mean that it is only the newer machines, the Nimbus and Archimedes, (and IBM PCs and Apple Macintoshes) which can accommodate them. DTP packages allow text to be edited to a much greater extent than a standard wordprocessor package. The latter allow text to be emphasised by the use of, for example, underlining, italics or bold lettering; but the former will also allow different fonts to be used at will, and in a range of sizes. Text manipulation packages are also available, (eg *FontFX* for the Archimedes) which allow text to be presented in a whole variety of ways (e.g. sloping in various directions, bent in an arc or a wave, with shadows on 'floor' or 'wall'). Text can be placed in boxes with various backgrounds and borders, and moved around with even greater facility than in most word-processor packages – the 'cut and paste' operations familiar to most teachers when producing worksheets can be performed on the screen, with the considerable advantage that the joins do not show! Packages are also available to enable data, derived from spreadsheets, databases or experiments, to be presented in a whole variety of ways (e.g. *Chartwell* or *Presenter GTI* for the Archimedes). Pictures can be incorporated with ease. These may be produced by the 'writer', using painting or drawing packages, which complement the DTP package; they may be 'imported' from libraries of drawings or pictures; or increasingly they can be produced using 'scanners' or cameras such as the Canon Ion camera. In principle, the pictures can be changed in size to suit the wishes of the author, and, although there are some restrictions on this, in practice there are few. A detailed account of these facilities would be out of place here, but it will be clear that DTP packages open the way to the production of very high quality materials both by teachers, and, in principle, by pupils.

If it were the policy of the school to attempt to provide the majority of pupils with DTP skills, then DTP could be used by the science department in just the kinds of ways suggested for wordprocessing above. Certainly, some pupils are sufficiently motivated by the ability to produce high quality material to make the extra effort to learn the necessary skills, and then do so, as usual, at a rate which most adults are quite unable to match! It is perhaps more realistic, however, to regard DTP as being more the province of teachers than pupils. The way is then open for a Science department to produce excellent materials, such as worksheets, or labels for displays – even whole displays themselves! (*Poster*, for the Archimedes, allows this activity) Figure 12 is an attempt to show a number of these aspects.

It will be apparent that it takes some time to learn the skills which are necessary to use all these facilities, and this is likely to be more than can be adequately compensated for by the satisfaction of producing pleasing materials. However, once the originals of worksheets are produced, the inevitable revisions then become relatively easy, since it is now a matter of editing and reprinting, rather than producing the whole worksheet from scratch again.

2 Less Established Areas

The aspects of New Technology discussed in this section are predominantly aspects of computer use and some of these aspects may well never become established in science education. Less space has therefore been devoted to them, but it has been thought worth mentioning even rather speculative ideas, because of the tendency for the availability of ideas and equipment to generate other ideas. No doubt, during the next decade there will be some quite novel applications in science education, which are not featured here!

Desk Top Publishing
IN SCIENCE

This page is an attempt to show the sort of effects which are provided by Desk Top Publishing packages, such as *Impression, Ovation* or *Pagemaker* and which can be used to produce worksheets and other written materials. **It must be said immediately that the impression may not be good because too many of these effects have been used.** Thus this paragraph has featured normal , italic and bold text, and a variety of font sizes. The heading also features different fonts, which have been manipulated using FontFX.

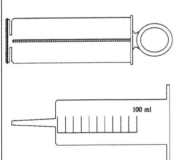

Pictures and diagrams are easy to import. The penguin is a piece of so-called 'clip art', while the other two diagrams have been drawn in the computer. The syringe is one small section of the extensive graphics library published by SSERC in Scotland. Such libraries make it relatively easy to construct diagrams from ready-drawn components.

100 ml

The diagram on the right is of a home-made colorimeter which can be used, when interfaced to a computer, to investigate the rate of the reaction in which sulphur is precipitated from a sodium thiosulphate solution. This diagram necessarily has to be drawn from scratch, since its subject matter is so idiosyncratic that no library is likely to be able to help in its construction. The size of this, and the other diagrams, has been easily adjusted to allow for the amount of text. This diagram has also been provided with a simple border.

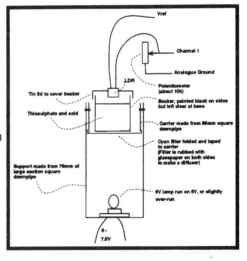

Figure 12 Desk top publishing in science

2.1 Hypermedia

These are packages which are rather similar to databases in some ways, but which have some novel features. An example, a hypermedia program on the Solar system, may make the basic principles clear. The introductory page might be as shown in figure 13.

Associated with each of the elements of the diagram is an area, called a 'button'. If the mouse arrow is used to select any particular button, say Jupiter's label, the program moves on to another page which gives further information about the planet. This, in turn, might contain text, pictures, diagrams or even music (perhaps the Planets Suite!) related to Jupiter. At various points on this page there will be buttons which will allow more detailed study of the aspect which the user wishes to follow up. This could be the history of the discovery of knowledge about Jupiter; information about or data on the planet or its moons; or information about the Great Red

Spot. The user can explore the databank more and more deeply, until there is no more; or alternatively return at any stage to the introductory page.

The power of the system is summarised in figure 14.

2.2 Multimedia

Relatively little has been produced hitherto in science, which uses a variety of media in a coordinated way but Resource have developed a package on *The Earth in Space* which contains a database, a simulation and a set of worksheets. However, in other areas of the curriculum, packages have been produced which use a computer program to show maps of an area, at a variety of scales, as the user 'travels' in search of information, which is provided by videotapes, audio tapes, slides or books. As the number of ways of storing data in the form of pictures, words or numbers increases, there is considerable potential for the development of learning packages in science which can draw on a variety of media.

---

1. The search among the information is in the hands of the user, and need not therefore be linear, or the same for each user;

2. the information may be in a variety of forms, not just text;

3. it can be used in 'reception mode', as described above, using ready prepared materials. Some fairly high-level text materials are available for use like this, but there is little or no commercially available material of a more varied kind, suitable for use with younger pupils;

4. or as with many computer packages, it can also be used in 'transmission mode' to enable pupils each to research a part of an area, such as the Solar system, and enter their data as part of a class production. This is essentially the same technique as that described in section 1 for using word processing in science education; and it has been used successfully in a school in the West Midlands.

Figure 14 Hypermedia: The power of the technique

2.3 Games

Pupils at all levels of attainment, including many who seem to find learning science difficult, seem to be able to learn computer games which appear incredibly complex; moreover, they are motivated to spend vast amounts of time playing them. Whilst it might be going too far to wish that it would be possible to package the whole of the science curriculum in the form of computer games, it would seem sensible to explore the possibility of tapping this source of motivation! Hitherto the use of games, on computers, in science education has not been common. Some simulations could be regarded as placing the pupils in the position of players, who are in control of the game, but most could hardly be regarded as fun! For instance, a common simulation is that of a nuclear power station, and all versions allow pupils to 'drive' the power station. What they will typically not do (particularly if the production of the simulation is sponsored by UKAEA!) is to allow the reactor to run away; nor do they introduce the essential element of competition, for instance by trying to run the station economically. Some simulations, however, do have such an element, but they tend to be complex, and since they have been designed to run on the BBC machine which has limited power in today's terms, they tend to be very slow. It could be that some of the simulations, rewritten for today's powerful machines, could utilise the motivating effect of the 'fun' element.

Many children are fascinated by the kind of game known as an Adventure game. While these exist on the commercial market, few do so for education. One that is available in science education is *Teletom,* which requires pupils to route a telephone message from the UK to the USA, by changing its form, using suitable transducers, to suit the variety of transmission media which have to be used. The classical adventure game format is not very evident in the simplest level of use, but at higher levels, restrictions may be introduced, such as random terrestrial or solar winds, which will affect satellite dishes and transmissions respectively, and lead to the need to modify the message routing. At the lowest level, children learn the terminology of communication and something of the processes involved, while at the higher levels the learning is much more detailed.

Games which test motor co-ordination are popular with both children and adults. There is a computer version of the old challenge of trying to thread a metal loop along an irregularly shaped wire without making contact and hence ringing a bell. As in the original the process of motor co-ordination and feedback can be discussed, and the effect of practice on performance investigated. However, the more important aspect of the game is that it allows exploration of the body's adjustment to a change in the feedback loop, produced either by inverting the action of the potentiometer or by putting a delay in it. The latter simulates the effects of abuse of alcohol, solvents or drugs.

2.4 Non-datalogging interfacing

There are a number of applications in which a computer is interfaced to an 'experiment', but with an aim other than to merely log data with time in a rather 'passive' way. In some cases there is a datalogging element, for example in plotting distance-time or speed-time graphs in real time, using either a commercial sensor which uses ultrasonic range-finding, or a 'Nuffield' dynamics trolley equipped with a contact, and running along a simple linear potentiometer made of conducting paper or wire. But the experiment is not simply datalogging and involves interaction between the user and some 'target' graphs, which are matched by moving oneself, or the trolley. This then provides an elegant way of learning about these graphs or testing knowledge of them.

Another application also involves interfacing a potentiometer with the computer in order to allow two pupils to work at a time, one setting a reading on an instrument, such as an ammeter, voltmeter, measuring cylinder, thermometer, burette or syringe. The other pupil has to read the instrument and enter the reading. The computer determines whether the answer is correct within a range, predetermined by the teacher or the pupils.

All of these applications are described in detail in *Logging on to Science* (Tebbutt 1991), as are other non-datalogging applications, such as a realistic simulation of the Millikan experiment; the use of the computer as a large digital ammeter, voltmeter, wattmeter, ohmmeter, *pH* meter, or thermometer; plotting *pH* curves with a true volume axis; using the computer as a human joulemeter or wattmeter, or as a digital timer.

A quite different aspect of interfacing is the use of a device called a *concept keyboard*. This was originally developed for special needs, for instance to help pupils who find the operation of a normal keyboard to be beyond them. The concept keyboard uses a membrane instead of keys, and can have areas designated as giving a range of responses, which are shown on an overlay produced specifically for each application. Since the concept keyboard offers only an alternative method of communicating a response, it can be used with a variety of programs. It does, however, need a program which 'tells' the computer which of the areas of the keyboard are to be inter-preted as particular responses. Such programs are *Touch Explorer Plus* for the BBC micro, or *Concept Designer* for the Archimedes.

2.5 Interactive video

Interactive video has been available for a number of years. The first such appli-cation was the *Domesday project* which combined data, photographs and maps of the British Isles, and which was therefore not specific to science. In order to be able to access this data at will it had to be stored on a large videodisc, which required a player; the computer which controlled the system needed special hard-ware. These extra requirements also applied to the first science disc which had data, pictures and film of various moving objects. The motion of the objects could be analyzed by the computer interacting with the videodisc. This was an exciting and powerful application of computing and video technologies, but it was expensive, particularly since the system was different from the Domesday system. Recently, three discs have been produced under sponsorship of the Nuclear Forum on *Radiation, Energy, and Risk*. Some parts of the system are the same as the Domesday one, but those specific to the computer, and the computer itself, are different. The various discs are not easily interchangeable, if at all, so the potential of the systems is unlikely to be realised until sufficient discs are available for one particular system in a range of school subjects, with enough in each to make it worth teachers' time in learning how to use the system.

2.6 CD ROM

This is currently (1992) new technology, which is similar to interactive video in some ways. It uses Compact Discs (CDs) instead of the large videodiscs, but is still able to store vast numbers of images, such as those of the outer planets and their moons, taken from spacecraft. Once again a player is neces-sary, together with a computer with an interface, and appropriate software, which makes the technology rather expensive, but there seems to have been more progress in establishing a standard, so that provided that enough discs are available, this technology may have more potential than videodiscs.

2.7 My world

This is a single program, for use with the Archimedes, which was originally intended for use with infants, or special needs pupils, and which does not seem, therefore to fit well in any section to do with science. However, its potential may be made clear by examining figure 15.

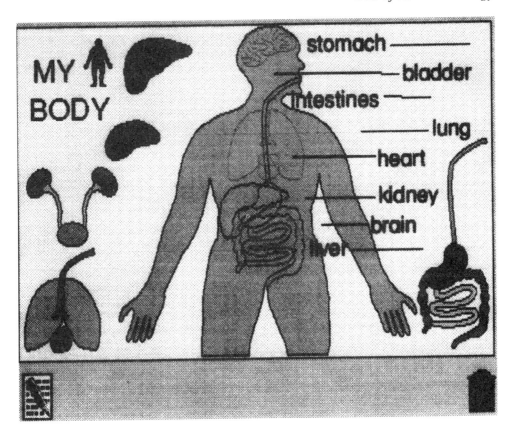

Figure 15 Using the program "My World"

This shows a screen (actually designed and produced by one of my students only 3 weeks after first learning to use the !Draw application) which portrays an outline of the human body, together with a selection of organs, arranged around the outside of the display. The *My World* program allows the organs to be moved, using the mouse, into the position which the pupil thinks is correct. Once the diagram is complete, it can be printed for later placement in the pupil's book. Screens can be prepared with diagrams and/or text, and pupils can add their own text if required. Different levels of complexity are possible. The program is cheap and easy to use with ready prepared screens with younger, or less able pupils. Since prepared Science screens are not currently available in quantity, it is necessary for the moment to prepare your own, which requires some skill, and more time, though the documentation explains how to do this in detail. It is to be hoped that a sufficiently large library of screens will become available to encourage the use of this program, without the need for users to have to design their own screens.

2.8 Control

The use of computers for control is an aspect of their use in the 'real' world which is increasing rapidly. This is because they can perform repetitive tasks with increasing precision, on the basis of instructions delivered by a program, or 'learned' by the machine, and modified as a result of measurements taken by the machine. The basic techniques have been a part of school curricula for much of the last decade, but have usually been dealt with by CDT departments. With increasing interest in biotechnology, and the relevance of technology to Economic and Industrial Understanding, it may be that control will feature more strongly within science than it has previously, though it is difficult to see how sufficient time might be found for this.

Control has been approached successfully via the ubiquitous 'Lego' system, or via extensions of the popular MFA system.

2.9 Modelling

Modelling and simulation are processes which have considerable similarities, but inevitably some differences too. Both depend on the ability to specify a mathematical model to describe some process. In a simulation the model is hidden from the user, though some of its characteristics may become evident as a result of its use. In contrast, and not surprisingly, models feature prominently in the modelling process. Their purpose is to allow the user to specify a model, and to see if its behaviour is like that of the original process, and to modify the model to improve it. The intention is to lead to greater understanding of the system which is being modelled, and also of the theory which is drawn upon in developing the model.

One approach to modelling, via spreadsheets, has been described above (p 157). The process of modelling, using specific software, tends to be done at the upper levels of schooling, and successful examples have been the *Dynamic Modelling System* for BBC and Nimbus machines, and more recently *Numerator* for the Nimbus and Archimedes machines. The Modus project is also developing modelling techniques for PC computers.

2.10 Weather satellites

Modern computers have sufficient computing power and memory to be able to process and display the transmissions of the various weather satellites. It is therefore quite easy for schools to set up a weather satellite station, which can lead to productive cooperation between departments such as Science, Geography and CDT.

3 Practicalities

3.1 Organisation of hardware

The new technology devices which have been mentioned in sections 1 and 2 are video recorders (with their monitors), computers (of course!, with their accessories), photocopiers, and players for videodiscs and CD ROMs. There is a considerable difference in the availability of the first three in this list from the remainder, and there is a similar difference in the range of uses of the devices. These points have implications for the way the devices are used and for the numbers which are required. The limited number of applications for videodiscs or CD ROMs means that at present few are needed. However, although the applications of the photocopier are many, they do not require that the machine is used 'in real time' or in class, so a machine may be located in a science department prep room, or as part of a centralised school reprographics facility.

Only videotapes and computers need to be used both in real time and in class, and the ways in which this can be done depend on the numbers which are made available.

It is quite clear that if teachers have to organise the provision of a video-tape recorder and monitor then they are likely to wish to make considerable use of it during that lesson. Moreover, they are only likely to make use of videos on those occasions, and not as frequently as they should. Most secondary schools have long since gone beyond the possession of only one video/monitor, but still fall short of having such a system in each teaching room. The extent to which videos can be used in lessons will depend on how easy it is to obtain a playback system. If one is shared between, say, 3 laboratories on the same floor, then teachers should be able to make quite effective use of videos, using small sections, as required, either for whole class use, or for use by small groups of pupils. On the other hand, if the facilities are centralised, the physical effort and time necessary to transport a system to the Science Department is likely to inhibit its use, even if similar total numbers of systems are available. The ideal situation is to have a video player (which is cheaper, and less attractive to thieves than a video recorder) and monitor permanently fixed in each laboratory. An LEA in the West Midlands is currently including such an arrangement in laboratories as they are refurbished.

The situation which applies to computers is similar, but at the same time more complicated. If laboratories do not have easy access to at least one computer, then computing is not likely to feature frequently in lessons. One way to ensure this is to have a computer permanently mounted in each laboratory. It can then be used for demonstration, or by groups as part of a circus, or as a general resource. This solution is being adopted by the LEA mentioned above.

This solution does, however, have problems. A large number of computers are fixed in position, and cannot be used together if required. An alternative way of organising the use of 'one computer per laboratory' would be to have each one mounted with its monitor and accessories (for example disc drives, buffer boxes and printers) on a trolley, in such a way that all that is necessary to use it is to plug in one lead to a suitable mains socket. However, this will only work if the laboratories are on one level, or access between levels is easy. If this is not the case, laboratories are usually grouped in threes or fours, and at the rate of one or two machines per laboratory, there will be enough machines within easy reach to do some serious work.

The trend towards having all the computers in a school located in one room, under the control of a computing department, seems fortunately, to have been reversed. Computer rooms may be whole-school resources, and/or only part of the computing provision in a school. In these circumstances the Science Department can do work on spreadsheets, wordprocessing, and simulations, as well as drawing on facilities such as CD ROM or weather satellite receivers which may only be available as a centralised resource. Unless the department has a number of dataloggers which can operate independently of computers, it is still likely to need its own computers for this purpose, since science experiments which are potentially wet, corrosive or alive are usually not welcome in computer rooms, even if there is room for them!

3.2 Software

Software is a problem, or perhaps, realistically, more than one! For computers, the software problems are finding what is available for your machine and managing its use in school.

Figure 16 shows a small illustrative sample of software, organised in the same sections as is this chapter.

A more comprehensive listing is readily available in either a catalogue from a general software supplier, or a software directory. Examples of both are given in the references. Software reviews in *School Science Review* and tours of the stands at the ASE annual conference are also useful sources of information, particularly about software which is sponsored by industry.

Once you have obtained your selected software you cannot assume that you can make copies for use with each computer. Most suppliers encourage

Area of work	BBC Machines	Archimedes	Nimbus/ PC
Datalogging	SCOPE (T) DATADISC+ (P)	SCOPE (T) DATADISC+ (P) INSIGHT (L)	SCOPE (T) DATADISC+ (P)
Spreadsheets	GRASSHOPPER (N) VIEWSHEET (A)	GRASSHOPPER (N) SCHEMA () EUREKA (L) DATASHEET (K)	GRASSHOPPER (N) WORKS (M) EXCEL (M) LOTUS 1-2-3 (LO)
Databases	KEY (I) QUEST (AU)	KEY+ (I) QUEST (AU)	KEY+ (I) WORKS (M) LOTUS 1-2-3 (LO) QUEST (AU)
Wordprocessing	VIEW (A) WORDWISE (C)	1ST WORD PLUS (A) PENDOWN (L)	WORD (M) WORKS (M)
Desktop Publishing		IMPRESSION (C) OVATION (B)	PAGEMAKER (AL)
Hypermedia		GENESIS (O) MAGPIE (L)	LINKWAY (IBM)
Games	TELETOM (BT)	TELETOM (BT)	
Non-datalogging interfacing	METER (T) LEARNER (T)	METER (T) LEARNER (T)	METER (T) LEARNER (T)
My World		MY WORLD (S)	
Control	LEGO LINES (LT)		
Modelling	DYNAMIC MODELLING SYSTEM (L)	NUMERATOR (L)	

Suppliers
- A = Acorn computers
- AL= Aldus
- AU = Advisory Unit (Herts)
- B = Beebug or RISC developments
- BT= British Telecom
- C = Computer Concepts
- I = ITV
- IBM = IBM
- K = Kudlian Soft
- L = Longman Logotron
- LO = Lotus
- LT=Lego Technics
- M = Microsoft
- N = Newman College
- O = Oak solutions
- P = Philip Harris
- S = North Western SEMERC
- T = M J Tebbutt

Figure 16 A sample of software

you to make a single 'backup' copy to use, while keeping the 'master' safe, but you are generally not allowed to make more copies than this unless you possess a licence to do so, either by having purchased it explicitly, or by being granted an implicit licence within the original purchase price. It is important, therefore, to have some system for maintaining the security of the master discs, and for controlling the copying of discs.

Acknowledgements
I happily acknowledge the work of two of my students, Lucy Harris and Martin Laslett, upon which figures 13 and 15 respectively have been based.

Maurice Tebbutt is a Senior Lecturer in Science Education at the University of Birmingham School of Education. He has been chair of ASE West Midlands and an ASE Council Member. He has a longstanding interest in computers in science education.

References

Acorn (1992) *Acorn Education Directory,* Acorn Computers Limited. A wide range of software, but for Acorn computers only.

AVP, *The Big Black Catalogue.* Useful because it lists much of the software available for the computers in greatest use in schools.

Flavell, J H and Tebbutt, M J (forthcoming) *Spreadsheets in Science: An introduction,* with numerous examples.

NCET (1992) *Educational Software,* J Whitaker and Sons Ltd. This book is subtitled 'A directory of currently available software for primary and secondary education' and no more needs to be said.

Tebbutt, M J (1991) *Logging on to Science,* ASE, Hatfield. Intended to be a comprehensive guide for those who are unfamiliar with 'interfacing'.

9 Use of Living Organisms

Roger Lock

1 Why use Living Organisms?

The biological sciences are one of the central components of science education today and there is a logic which links the use of organisms in science lessons to biology, the study of life. Modern approaches to teaching emphasise a pupil-centred approach, with a focus on direct, first-hand, experience of the world of science, a world that includes organisms, living and once lived. The central tenet of the association between the study of biology and first hand experience is recognised by the major organisations concerned with science education; the Association for Science Education (ASE) the Institute of Biology (IOB) the Royal Society (RS) and the Universities' Federation for Animal Welfare (UFAW) (RS/IOB, 1975; ASE, IOB, UFAW, 1984). Such links are endorsed and explicitly identified in the *Science National Curriculum* (DES, 1991). For example, the programme of study for Key Stage 4 (Sc2) requires that

> *Pupils should make a (more) detailed and quantitative study of a locality, including the investigation of the abundance and distribution of common species, and ways in which they are adapted in their location.*

The links were supported by the Department of Education and Science (DES), now the Department for Education (DFE), who in a memorandum to Local Education Authorities (DES, 1989) stated with respect to animals:

> *There are many excellent reasons for introducing live animals into the school environment; animals are not found in schools as often as they could be and there are sound educational advantages to be gained from pupils having immediate experience of animals right from the early years.*

There is not space here to explore the many reasons for using organisms in school science lessons; such issues have been covered elsewhere (Lock, 1992, 1993). The emphasis of this chapter will be upon giving examples of what organisms may be used, where and how.

In trying to write succinctly on this topic, it is inevitable that some specific, biological knowledge has to be assumed. The non-biologist may require help from a specialist colleague in interpreting some of the details, a situation now common in a balanced science department. At many such points reference is made in the bibliography to more detailed resources.

It would be wrong to leave this section without a brief consideration of the responsibilities of science teachers to their pupils' feelings. Pupils hold clear opinions about the uses to which organisms are put in school science (Millett and Lock, 1992) and, while such attitudes may not always be reflected in behaviour, or founded on an accurate or extensive knowledge base, there is a need to recognise and respect them, to tell pupils about the impending use of such material, to articulate the aims that such work intends to convey and to give pupils the opportunity of alternative comparable activities if necessary.

There is a further obligation to draw to pupils' attention the kinds of ethical and moral issues that are faced by scientists in their work relating to the use of living organisms. For example, the National Curriculum (Sc2, Level 10) provides a context where the potential conflict between management systems used in farming and animal welfare should be considered. Above all, in our work with living organisms, we should aim for aware and informed pupils who can weigh evidence, distinguish between fact and hearsay, and make up their own minds about the way in which society uses living organisms, from the pet cat to the laboratory mouse, the racehorse to the intensively-reared rainbow trout and from the genetically engineered microbe to the Amazonian rain forest.

2 What Sort of Living Organisms?

Although there is now a five kingdom system of classification, for the purposes of this chapter three major groupings are used as follows

- Microbes: Fungi, Bacteria, Proctista (Protozoa, Algae)
- Plants
- Animals

The ensuing sections of this chapter are organised under these three main headings and are followed by a section on Living Things and the Law and a select bibliography, complete with some useful addresses.

3 Microbes

The essential reference for all work using microbes is the HMI guide. (DES, 1990). Much of the information in this section draws on advice contained in this text.

3.1 Why teach about microbes?

Microbes are very important because they are:

- of medical importance e.g produce human insulin for diabetics
- of industrial and economic importance e.g produce acetic acid and citric acid
- of social importance e.g in public and domestic hygiene
- involved in food production e.g mycoprotein, bread, alcohol
- involved in disease e.g cholera, influenza
- involved in cyclical changes e.g decay of dead organisms, sewage disposal, nitrogen cycle.

They also have a role in the formation of mineral deposits (oil) and in the digestion of herbivorous animals.

For science teachers there are additional reasons for using microbes, such as their usefulness in investigating some of the properties of living organisms (the need for a food source, growth, reproduction, respiration, excretion) and in exploring a range of biological principles and processes, eg fermentation, growth curves, mutation, and the properties of enzymes.

3.2 Starting points for work on microbes

Most pupils at 11 years of age will have experience of moulds and mildews; they may know about yeast or have seen rusts without realising they are microbes. They will know about germs and about the symptoms of illnesses or diseases caused by bacteria and viruses. In addition, they will have seen "green stuff" on trees, walls or branches, without perhaps appreciating the link to unicellular algal colonies.

Early work could involve observing some of these microbes as shown in table 1.

Table 1 Observation of microbes

Macroscopically plus using a hand lens	{rotting fruit (sealed in a container) {rotting wood {rotting leaves/leaves with rusts
With a binocular microscope and/or hand lens	{bacterial and fungal colonies {growing on agar in sealed {petri dishes
Microscopic examination of wet mounts made by the pupils	{protozoa – from pond mud or hay infusion {diatoms – collected from freshwater {yeast – grown in school {algae – scraped from tree bark

Such work makes pupils aware of the microbes' existence and of some of the variety, and is beginning to enable them to take tentative steps towards distinguishing between them. An example, for bacterial and fungal colonies grown in taped-down petri dishes, is given Table 2.

Of course, not all bacteria or fungi conform to such patterns, but the table provides clues to help with identification on the basis of which column of criteria offers the 'best fit'.

3.3 What should pupils know by the end of KS4?
By the age of 16 all pupils should have the basic knowledge of microbes shown in Table 3.

3.4 What sort of practical activities can be done by pupils?
Many schools have avoided practical work in microbiology thinking it is illegal or unwise. It is *not*. Care should be taken to follow the guidelines provided by the ASE (1988) and the DES (1990). Given such sensible practice, all the work described below, and that given in resources in the appendices, can be successfully and safely carried out.

For reasons of space, the work mentioned will be limited to that which aims to supplement and illustrate the knowledge and understanding identified in Table 1.

Table 2 Distinguishing between types of microbe

	BACTERIA	FUNGI
Colony Size (seen from above)	Small 1mm to 10mm diameter	Large may fill the whole dish
Height (seen from the side)	Flat	Tall
Appearance	Surface is shiny	Surface is dull
Colour	Yellow, Pink, it varies	Grey, White (spores are coloured sometimes)
Texture	Like a drop of liquid	Like cotton wool
Growth medium	Grow best on nutrient agar (more acid)	Grow best on malt agar (more alkaline)

Table 3 Knowledge and understanding that all pupils should have of microbes by 16 years of age (Adapted from DES (1990))

- Microbes are a varied group
- Microbes are widespread
- Some conditions make them grow rapidly
- Other conditions inhibit their growth
- Some microbes are helpful in food production in industry, in medicine
- Other microbes cause disease and illness
- Microbes are essential to the balance of nature
- Microbes must be handled and disposed of safely
- Microbes can mutate and change their characteristics and this ability is used by humans in the production of eg. antibiotics, enzymes, hormones, animo acids.

The observation work described on page 182 goes some way to showing that they are a varied group. Such work could be combined with investigations, using prepared petri dishes that set out to answer questions such as:

- Do microbes grow on leaves?
- Are there microbes in soil?
- Do microbes live in water?
- How many bacterial spores land on a petri dish exposed to the air for 60 seconds? 30 minutes?
- What grows best on nutrient/malt agar?
- Does the dishwasher in the canteen leave plates more free of microbes than if they are washed by hand and dried using a tea towel?

The dishes used in such investigations need to be sealed after exposure and labelled as shown in Figure 1. Petri dishes are designed so that gases can diffuse in and out. Plates should not be sealed around the rim with tape as this will cause an atmosphere depleted of oxygen inside the dish. In these conditions you are selecting for the growth of anaerobic microbes, many of which arc pathogenic (disease causing). You can make a complete cross with the tape if necessary, but this may obscure vision.

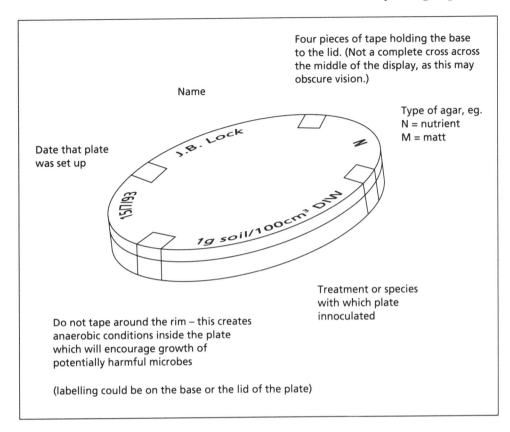

Figure 1 Taping a petri dish and labelling it for later examination by students

Plates should remain sealed during examination by pupils and through to disposal. Pupils can set up such practical work for themselves and their findings should help to lead them to an understanding that microbes are widespread, and to some perception of conditions that inhibit and enhance their growth rates: further work could challenge pupils to produce, for yeast, the fastest rate of evolution of carbon dioxide, or the densest culture (this usually being achieved in aerobic conditions). As well as varying the oxygen concentration, pupils could change temperature, *pH*, light and the type of substrate and its concentration, and add nutrients such as phosphate, nitrate, black treacle or coconut milk.

Other investigations could involve the role of yeast in raising dough, and changes in the viscosity of milk on addition of yoghurt-making bacteria. Such work could illustrate the role of microbes in food production as well as exploring conditions influencing growth. For practical work on microbes in industry see Dunkerton and Lock (1989) and Henderson and Knutton (1990).

Investigations involving the "damping off" of cross seedlings give insights into the role of microbes in disease, while the use of sensitivity discs (discs impregnated with antibiotics), allows the influence of antibiotics, of a variety of different types and concentrations, on bacterial growth to be explored. It is possible to explore the claims of manufacturers of disinfectants, antiseptics, deodorants and toothpastes by making your own discs impregnated with these substances at differing concentrations and introducing them onto the surface of a petri dish inoculated with a culture of a safe bacterium such as *Bacillus subtilis*. Discs can be made from filter paper, using a paper punch, or wells can be cut in a plate using sterile cork borers. (Figure 2)

Practical work exploring the role of microbes in recycling materials could be carried out by investigating the biodegradation of supermarket plastic bags. Safe handling and disposal could be demonstrated by the teacher and/or technician.

No attempts should be made in school to induce bacterial mutation or to practice the techniques of genetic engineering (DES, 1990).

3.5 Pupil safety rules

Pupils in KS3 and 4 who are doing work with microbes need to be reminded of the safety rules that apply.

3.6 What are the advantages of using microbes?

It is a good idea to discuss with pupils the reasons why microbes are used in work in biology.

3.7 The contribution of microbes to Biotechnology

There has been a tendency to merge the terms microbiology and biotechnology in some school science contexts. It is important to introduce pupils to the significant contribution that microbes make to food production, the development and production of industrial and medical products, and the microbial role in the treatment of waste matter. They should realise there are elements of biotechnology in which plants and animals are involved.

For further work on the role of microbes in biotechnology see the Resources section.

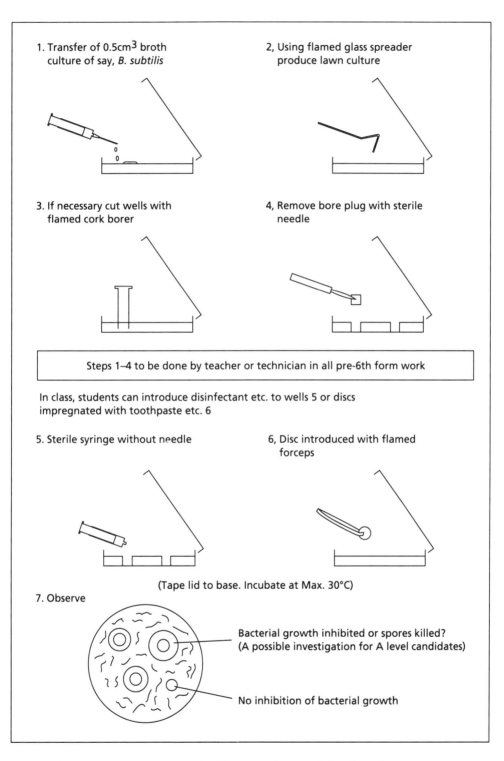

1. Transfer of 0.5cm³ broth culture of say, *B. subtilis*

2, Using flamed glass spreader produce lawn culture

3. If necessary cut wells with flamed cork borer

4, Remove bore plug with sterile needle

Steps 1–4 to be done by teacher or technician in all pre-6th form work

In class, students can introduce disinfectant etc. to wells 5 or discs impregnated with toothpaste etc. 6

5. Sterile syringe without needle

6, Disc introduced with flamed forceps

(Tape lid to base. Incubate at Max. 30°C)

7. Observe

Bacterial growth inhibited or spores killed? (A possible investigation for A level candidates)

No inhibition of bacterial growth

Figure 2 Investigating control of bacterial growth by disinfectants, etc

187

4 Plants

4.1 But there's not much room for plants at school and they're too much trouble to keep!

If you see this as a problem, then worry no more. In the following paragraphs and table 4 five plants are described that are indispensable, with which you can cover practical activities for most botanical topics. Two of them, *Elodea sp.* and *Lemna sp.*, require little or no care and attention, as they can be simply placed in your school pond (or an old sink or bath if you don't have a pond).

The three land plants Geranium, Busy Lizzy and *Tradescantia sp.* require little maintenance and can be successfully kept on laboratory window sills, in a Wardian window or Greenhouse (although they need to be protected from frost in the winter).

To get investigations using the evolution of bubbles from the broken ends of *Elodea sp.* stems to work well ((c) and (f) in table 4), the plants need to be brought in from the pond to an aquarium and continuously, brightly illuminated and aerated for at least 48 hours before use in class. The traditional demonstration which collects the gas evolved using an inverted filter funnel and a boiling tube, serves as a suitable investigation for A level students who could be encouraged to analyse the gas collected. Many pupils (and some teachers too!) are surprised to find that it is not 100% oxygen. How did that teacher demonstration work so well when you were at school?

The land plants can be maintained by taking cuttings, a task that not only teaches pupils valuable skills, but also has wider applications for future citizens with an interest in gardening, not to mention providing a means of raising additional income at the school fete.

4.2 What about fast plants?

In recent years, the rapid cycling *Brassica rapa* has been introduced to this country from America, which completes its life cycle in five weeks: yes five weeks from seed to seed! The plants are small and relatively easy to maintain but they do need a light bank of incandescent tubes to maintain them successfully. They are excellent for investigations into germination and growth, pollination and embryology. Leaf discs can be used in photosynthesis experiments. A short in-service course (one day or less) is advisable before undertaking work with these exciting plants and further details can be obtained from Richard Price, Director of the SAPS Project at Homerton College, Cambridge (See Resources).

Table 4 Five plants for science teaching

Plant name	Uses
Elodea canadensis *Elodea densa* Canadian Pond Weed	a) Structure of plant cells (leaf is one cell thick) b) Chloroplasts circulate in leaf cells in response to damage (cyclosis) c) Elodea "bubbler" experiments where factors affecting the rate of photosynthesis are investigated d) Gas exchange in photosynthesis and respiration by immersion in bicarbonate indicator solution e) An example of asexual reproduction f) Collecting gas evolved from continuously illuminated plant (oxygen evolution in photosynthesis?)
Pelargonium zonale Geranium	a) Leaf prints to demonstrate the presence of starch (Photosynthesis experiments that show the role of light – mask a leaf, need for CO_2 – leaf in flask with KOH, need for chlorophyll – variegated leaf) b) Artificial propagation by stem cuttings c) Cultivars show continuous and discontinuous variation d) Translocation experiments (Transpiration) using potometers.
Impatiens balsamina *Impatiens wallerana* Busy Lizzy	a) Transpiration experiments using potometers b) Leaf replicas/leaf peels to investigate leaf surface structure c) Photosynthesis experiments (leaf prints as for geranium)
Tradescantia albiflora (a wide range of species and varieties) Wandering sailor	a) Investigating mineral requirements for growth (mineral culture) b) Photosynthesis experiments (leaf prints with variegated leaves) c) Protoplasmic streaming in staminal hairs d) Osmosis investigations with staminal hairs e) Growth of pollen tubes f) Continuous and discontinuous variation.
Lemna sp. Duck weed	a) Mineral requirements for growth (mineral culture) b) Growth rates (count the fronds) c) Asexual reproduction

4.3 Horses for courses – particular plants for particular topics

There are always particular points that need demonstrating or aspects of the biological sciences that pupils want to investigate where plants may be a useful resource. In Table 5 there is a range of plants that may be used for specific purposes.

4.4 Using the supermarket and greengrocer as a resource.

Many of the plants used for food are very valuable in the teaching of science. Table 6 shows some of the ways in which common plants found in super-markets can be used.

The information in Table 6 is not exhaustive and comprehensive. For example, no mention is made of the use of cut flowers or seeds. Supermarkets are very useful sources of information on food, nutrition, healthy eating and food additives. They also provide a valuable resource for use in scientifically based investigations into "organic" and other "green" products, ozone "friendly" and cholesterol free products, packaging and advertising, biodegradable boxes and bags and biotechnology in food production, to mention but a few.

5 Animals

Concise information on the culture and maintenance of a wide range of living organisms can be found in Archenhold, Jenkins and Wood-Robinson (1978) and in the CLEAPSS Laboratory Handbook (1989). Suffice it to say here, that science teachers have a range of responsibilities with respect to animals and their welfare. Teachers should be practised in handling and in care and main-tenance skills. In addition, they should recognise the health of their animals and know when to seek veterinary advice. Above all, they have a duty to show a caring and humane approach in all their work with living things.

Few issues in science education have aroused as much controversy in recent years as the use of animals, living and dead, in science lessons. Such controversy is part of a wider discussion about the ways in which society uses animals. Here there is only space to comment on those issues most central to the concerns of science teachers. These might be summarised as follows:

- animals as pets and responsible pet care
- animals used in science education
- social, ethical and environmental issues about the use of animals by scientists.

Table 5 **Plants for specific purposes** (after Tessa Carrick)

Investigation	*Useful plants*
Chromatography	a) Petals of geranium *(Pelargonium zonale)* especially red ones b) Leaves of *Coleus blume* c) Leaves of *Begonia rex*
Reproduction (Sexual)	a) Separate male and female flowers *Begonia rex* or *Begonia semperflorens* b) Shepherd's purse *(Capsella bursa-pastoris)* a common weed that can be found in flower at all times of the year. Ovules at various stages of development can be teased out from ovaries with 'A' level students
Vegetative Propagation	a) Spider plant *(Chlorophytum sp.)* b) Mother of thousands *(Saxifrage sarmentosa)* c) Chandelier plant *(Kalanchoe tubiflora)*
Convergent evolution	a) Variety of succulents
Adaptation to environment	a) Variety of succulents b) Variety of cacti
Genetics	a) "Genetic" tomatoes b) "Fast" plants *(Brassica rapa)*
Nastic reactions to touch (plants move fast!)	a) Sensitive plant *(Mimosa pudica)* b) Venus fly trap *(Dioneae muscipula)*
Insect "eating" plants	a) Venus fly trap *(Dioneae muscipula)* b) Sundew *(Drosera sp.)* c) Butterwort *(Pinguicula sp.)*
Fascination and entrancement	a) Amaryllis – rapid growth and beautiful flowers b) "Resurrection Plant" – dry, apparently dead, ball of material comes to life and can be dried out again c) "Dinosaur Plant" – as above and really a club moss – contributor to coal seams.

Table 6 Biology in the supermarket (modified after Tessa Carrick)

Plant	Plant use	
Onion	Epidermis peel (from swollen leaves in the bulb)	– plant – cell structure – stain the nucleus – osmosis experiment
	Roots Bulb structure	– cell division/mitosis
Rhubarb	Epidermal peel	– osmosis investigations, easy to see plasmolysis because of pigment in cell sap. – cell structure
	Petiole structure	
Sprout	Bud structure	– get pupils to be creative biotechnologists and design a "super plant" with sprouts for buds, rhubarb for petioles, lettuce for leaves etc.
Lettuce	Leaf Epidermal peel	– chromatography of chlorophyll – leaf structure – stomatal distribution
Celery	Petiole	– transpiration by dye uptake – vascular bundle structure
Potato	Stem	– osmosis investigations with chips/bored cores – vegetative propagation
Red cabbage	Leaf	– extract pigment, make your own indicator paper or liquid indicator
Cress	Whole plant	– germination – growth – root hairs
Tomato	Fruit	– seed germination – inhibition of germination
Apples	Fruit	– enzyme activity (browning of cut, exposed surfaces) – extraction of juice from pulped apple using pectinase (an industrial application in the fruit juice industry)
Radish	Root	– osmosis investigations into curvature of segments cut radially (decorative functions in catering industry)
Peas/Beans	Fruit (pods) Broad bean seeds	– fruit/ovule relationship – germination – tropism investigations
Cauliflower	Flower	– plant tissue culture
Carrots	Root	– root structure
Peppers	Fruits	– taste testing

The most significant abuse of animals in this country is, arguably, that offered to pets. There are aspects of responsible pet care in the National Curriculum, mainly in KS2. Recent research has shown that there is wide-spread pet ownership by pupils in KS4 (Lock and Millett, 1992a) and, conse-quently, there must be a place in the curriculum for raising issues of consid-erate ownership. In addition, there is an obligation on science teachers to illustrate humane care through the ways in which animals are used in school science lessons. Further resources are identified in the bibliography.

As science teaching is concerned with the social, ethical and environmen-tal issues faced by scientists and by society, the use of animals in scientific research, for example in biotechnology, genetic engineering and in the devel-opment of new medicines, is an aspect which merits consideration; indeed, the National Curriculum (Sc2 Level 10) makes such consideration explicit. The bibliography identifies resources that make such work possible.

5.1 Living animals in science lessons

When discussing the use of living animals in science lessons there may be preconceived ideas of an albino mouse in a cage. However, there are other possibilities to consider in addition to those animals, such as gerbils and locusts, which are maintained and bred in schools, out of their natural envi-ronment. Opportunities for work abound with organisms such as garden snails and slugs, woodlice and other terrestrial and aquatic Arthropods, which have been temporarily removed from their environment and which, when the observations and/or investigations are complete, will be returned there unharmed. In addition, much work can be done on behaviour, growth, population size, adaptation to environment, food chains etc, on animals in their natural environments without, temporarily, removing them.

Table 7 (pages 194-5) gives a selection of animals that are used in school and the uses to which they can be put.

Table 7 is arranged by phylum for convenience purposes only. Some phyla are unrepresented e.g Reptilia, Platyhelminthes and Nematoda, because they are not commonly used in school science contexts. The National Curriculum has 'an understanding of the variety of living organisms' as a clear require-ment for pupils in KS 2/3, and teachers may choose to use living examples of types from Phyla to illustrate the variety. Such a use has not been included in Table 7 as it could apply to all the animals listed. Clearly teachers would not want to miss the opportunity to illustrate the *Platyhelminth* phylum when a planarian was found while pond dipping, but other examples of that phylum,

Table 7 – Uses of Living Animals in Science Lessons

Animals	Use	

MAMMALS	Humans
	Homo sapiens
	(The most useful living
	resource in laboratories)
		– Continuous and discontinuous variation
		  – Hand span
		  – Shoe size
		  – Eye/hair colour etc
		– Ventilation rate/heart rate
		– Living organisms produce CO_2
		– Reaction times
		– Learning (mirror drawing)
		– Many others e.g vision, hearing, taste, touch, smell

	Gerbils/Mice *Meriones unguiculatus* *Mus musculus*	– Growth – Reproduction – Behaviour – feeding, grooming, exploratory behaviour – Genetics – more limited use – External structure – Living organisms produce CO_2
BIRDS	Zebra finch *Taeniopygia castanotis* Birds outside the laboratory	– Feeding behaviour – Synchronised behaviour in flocks – Food preference – Behaviour
AMPHIBIANS	African clawed toad *Xenopus laevis* Axolotyl	– Microscopic observation of capillary flow in web between toes. (No anaesthetic – hold firmly in damp cloth). – Locomotion – Adaptation to environment – Reproduction – Development
FISH	Gold fish	– Feeding behaviour/food preference – Variation – Locomotion
ARTHROPODS – CRUSTACEANS	Woodlice *Oniscus asellus or* *Porcellio scaber* Water flea *Daphnia*	– Behaviour – turn alternation – choice chamber experiments – Respiration – Heart beat rate and factors affecting – Peristalsis – Reproduction (young seen developing inside the female carapace) – Behaviour (bubble CO_2 into *Daphnia* in a water column, illuminate column at different points).

Animals	Use	
ARTHROPODS – INSECTS	Locust *Schistocerca gregaria* *or Locusta migratoria*	– Reproduction/behaviour – Growth rates (incomplete metamorphosis) – Feeding (use of mouthparts and food selection) – Respiration (use in respirometer and investigate factors affecting ventilation rate) – Movement (walking, jumping and flight)
	Fruitfly *Drosophila melanagaster*	– Observation of an organism that has been important in the understanding of genetics – Variation – Distinguish male and female – Behaviour (mating, selection of colour of background on which to rest) – Life cycle – Genetics ('A' level)
	Blue Bottle larvae (Maggots) *Calliphora sp.*	– Behaviour (choice chambers) – Movement (rate at different temperatures) – Movement in response to light direction – Growth and development (complete metamorphosis)
	Stick Insects *Carausius morosus*	– Behaviour (walking movement) – Reproduction (parthenogenesis) – Adaptations to escape predation (camouflage)
	Flour beetles *Tribolium confusium*	– Behaviour (choice chamber experiments e.g light/dark, humid/dry)
MOLLUSCS	Giant land snail *Achatina fulica* Garden snail *Helix aspersa*	– Food preference – Locomotion – Food preference – Locomotion
ANNELIDS	Earthworm *Lumbricus terrestris*	– External features – Behaviour (burrowing, feeding) – Movement
	Marine worm (tube living) *Pomatoceros triqueter*	– Fertilisation of living egg (mostly useful in 'A' level contexts, during field work)
COELENTER-ATES	*Hydra sp.*	– External features – Feeding

such as liver flukes and tape worms, and of the *Nematode* phylum, might be best studied as preserved specimens.

5.2 Using living animals in their environment

The National Curriculum is explicit about studying animals in local habitats (DES, 1991). In all such work, care must be taken not to exploit or destroy the habitats studied. Suitable ones can be created within the school grounds by measures such as allowing grass to grow, planting native tree species and "importing" some dead wood.

For further information on work in local habitats see chapter 17 and the bibliography.

5.3 Dead animals in science lessons

Included in this subsection is a consideration of the use of parts of dead animals as well as whole ones. There is not space to consider the use of stuffed animals, animals preserved as whole specimens, museum mounts, skeletons, joints, skulls or individual bones or the use of thin sections. Consideration should be given to the ethics of killing animals specifically for the purposes of stuffing them or making museum mounts and specimens as against, say, using animals that have died naturally, been destroyed humanely by vets or killed for human food consumption.

A range of alternatives to the use of animal material is available (see Bibliography) including audio visual aids, charts and models. Casts are obtainable of skulls and bones.

It is the view of the author that all of these are a useful supplement to, but not a substitute for, "the real thing". However, for those with moral and ethical objections, the study and investigation of surrogate materials is preferable to no study at all.

5.4.1 The dissection issue

Dissection is the cutting up of dead animals and plants. It is important to be clear that the animals, or parts of animals, involved are dead, because there has been confusion by pupils between dissection and vivisection. (Lock and Millett, 1992b). Dissection of plants, living, dying or dead, appears uncontentious to most people and is not discussed further.

An attempt is made to give a balanced view of the case for and against dissection in Table 8 (page 197) but reference to further resources on this issue is given in the Bibliography.

Table 8 Some arguments used for and against dissection

The case against	*The case for*
An education in care and compassion is more important than science education	Learning by first hand, personal involvement is more effective than by using surrogate materials
Little educative value in dissection	Intricacy and delicacy of tissues appreciated
Models/a.v. aids are more effective	Three dimensional relationship between tissues and organs appreciated
Teachers are poorly trained in animal care and animal husbandry	Respect for life enhanced through increased awareness of structure
Pupils have a right to refuse to dissect	Skill development e.g problem solving, observing, recording (not routine dissection to develop manipulative skills)
Dissection desensitises pupils to the welfare of animals	Vocational role in further/higher education and veterinary/medical career

Pupils should make their own minds up about the issue of dissection on the basis of balanced information: but an objection to dissection is not a reason for avoiding study of the biological sciences. It may be that some pupils object to the dissection of whole animals but accept the use of abattoir materials. In all cases, teachers are obliged to discuss this issue with their pupils and to support and protect both those that opt out and those that opt in from peer pressure.

5.4.2 Parts of dead animals
Subject to the wishes of pupils and appropriate locations in work-schemes, the study of parts of dead animals can occur in many places in secondary school work, mostly in KS4 and at A level. Some such potential uses are given in Table 9.

The last example given in Table 9 raises some interesting questions. Since the DES recommended that human epithelial cheek cells should not be col-

Table 9 Science in the butcher's and fishmonger's shops (modified from Tessa Carrick)

Part of Animal		Use
Chicken leg and foot		– muscles, tendons and joints
Pig trotters		– joint structure and function, cartilage, tendons, ligaments, muscles
Lungs		– muscle, skin, nerve tissue – gross structure of lung, trachea – location of heart with reference to lungs – dissection of bronchi and bronchioles – inflate lungs with pump – spongy nature of lung, floats in water
Heart (Health regulations lead to much damage – select your specimens with care)		– heart structure – valve structure/function – muscle and fat tissue
Kidney		– gross structure and internal (L.S.) structure – ureter, renal artery and vein, adrenal gland – cut thin slice, tease out individual nephron ('A' level)
Eye		– gross structure, external features – nerve, muscle, fat – dissection and experiments with lens
Bones (boiled first)	Long marrow bones	– cut in LS/TS for bone structure and nature of joints
	Vertebra (neck of lamb)	– bone structure and protection of spinal cord
Blood (citrated on collection and kept in the fridge)		– colour and texture when O_2 and CO_2 bubbled through. Reversible nature of these changes. Not reversible if CO bubbled through – volume of O_2 carried by solution ('A' level)
Fish	Sprats/whitebait	– external structure, gills/fins – muscles and movement
	Cod heads	– movement of jaws – gill structure , gill rakers separate food from water – gill filaments mount on slide – epithelial cell scraped from inside jaw

lected from pupils and used as representative of a "typical" animal cell, there has been a problem over a readily available source of living animal cells that can be used as a substitute. Denise Young of Enfield County School has suggested that transparent adhesive tape placed firmly on a sensitive area, such as the inside of the wrist, will, when removed, take with it epithelial cells with visible nuclei. My investigations have found the epithelia from inside the mouths of cod or other fish to be more suitable.

5.4.3 Whole dead animals

Most dissection of whole dead animals is for the purposes of investigating the anatomy and structure related to function of all body systems. For A level students, however, there is the possibility, with small mammals, of investigating the diaphragm function. Such students could also carry out chromatography of gut contents to investigate the sites of secretion of enzymes and absorption of the products of digestion. In addition, the variation in the blood sugar concentration in different vessels can be explored. With locusts, as well as the chromatography of gut tissue, investigation can be carried out of the structure of the tracheal system and of nerve/muscle physiology.

Dissection of whole dead animals, if considered necessary with KS3 or 4 pupils, should be carried out as a demonstration by the teacher.

6 Living Things and the Law

There is a lot of misinformation and misunderstanding about the legality of keeping and using organisms, both living and dead, in schools. In particular, if an animal, such as a mouse or a rat has been killed humanely it is not illegal for it to be used for dissection and physiological work on organs and tissues, with pupils for whom teachers consider this to be suitable work.

The most succinct source of information on the legal aspects of using living things in school is administrative memorandum no 1/89, *Animals and Plants in Schools: Legal Aspects* (DES, 1989). The great advantage of this resource is that it uses lay terms to explain the legislation. A copy of this document was sent to all schools.

The two most important Acts are the *Wildlife and Countryside Act, 1981* and the *Animals (Scientific Procedures) Act 1986*.

6.1 Wildlife and Countryside Act, 1981

A twelve page booklet, *Protecting Britain's Wildlife* (DOE 1988) gives an excellent brief guide to this legislation and the summarised information given in the following paragraphs is taken from that guide and from the DES memorandum 1/89 (DES, 1989)

- all wild birds, their eggs and nests (while in use) are protected.
- 82 species of bird (see DES, 1989) are specially protected and a licence (available from the Nature Conservancy Council) is required to approach or photograph these birds if on or near their nests.
- 30 species of other animals must not be injured or taken, even for ecological marking studies. These include all bats, red squirrels and otters as well as some reptiles, amphibians, fish, butterflies, moths, crickets, dragonflies, beetles, grasshoppers, spiders and snails.

Perhaps the most important pieces of legislation covered by the Act relating to teachers and field work are

- it is illegal to intentionally uproot any plant grown in the wild unless it is on your own land or you have obtained permission.
- for certain rare plants (see DOE, 1988; DES, 1989) it is illegal to pick, uproot, destroy or collect flowers or seeds from them.
- it is illegal to capture shrews without a licence. Longworth, or other traps can be made legal by drilling a 13 mm shrewescape hole in the side.

6.2 Animals (Scientific Procedures) Act, 1986

An animal in the terms of this act is a non-human vertebrate animal and its foetus, free living embryo or larva. The Act makes the following procedures illegal in schools.

- decerebration of an animal (e.g frog) by pithing.
- injection of hormones (e.g African Clawed Toad) unless for husbandry purposes.
- nutritional experiments involving restricted or excessive diets
- anaesthetising an animal (e.g goldfish) to observe capillary flow.
- examining a living chick embryo after the halfway point of incubation.

The act does not make it illegal to investigate heart or nerve/muscle physiology of amphibians in schools as this could be carried out, if teachers considered it a wise and acceptable procedure, after humane killing (Lock, 1989). It is important to consider whether procedures which are not illegal are accept-able. Some experiments with invertebrates may be considered unwise in that they offend pupil sensibility; for example, removing the antennae from woodlice. While there are no legal restrictions on carrying out such experi-ments, one should seriously question the wisdom of such a strategy.

Roger Lock has worked in schools in Scotland, Birmingham and Warwickshire and in the Science Education departments of the Universities of Leeds, Oxford and Birmingham. He has served on three ASE region committees. Most recently he has been director of the Animals and Science Education project.

Select Bibliography and Useful addresses

Microbes
ASE (1981) *Safety in school microbiology.* Education in Science, 92, 19-27.
Bishop O (1984) *Adventures with micro-organisms,* John Murray, London.
DES (1990) *Microbiology; an HMI guide for schools and Further Education.* Second impression. HMSO, London.
Fry P (1977) *Micro-organisms,* Schools Council/Hodder and Stoughton.

Protozoa/Algae
CCAP (1988) *Culture kit for practical microbiology and biotechnology: Guide for teachers.* Shell Education Service, Society of General Microbiology, British Phycological Society. Available from Culture Collection of Algae and Protozoa, Freshwater Biological Association, The Ferry House, Ambleside, Cumbria, LA22 0LP

Biotechnology
ASE (1988) *Biotechnology,* Education in Science, 126, 12-16.
Dunkerton J & Lock R (1989) *Biotechnology: A resource book for teachers,* ASE, Hatfield.
Henderson J & Knutton S (1990) *Biotechnology in schools: A handbook for teachers,* Open University Press, Buckinghamshire. This book is a rich source of references to other sources.

Sources of general information on microbes and biotechnology
CLEAPSS School Science Service. See page 147.
MISAC (Microbiology in Schools Advisory Committee)
The Institute of Biology, 20 Queensberry Place, London, SW7 2DZ
National Centre for School Biotechnology, Department of Microbiology, University of Reading, Reading, RG1 5AQ

Plants
Bingham C D (1977) *Plants*, Schools Council/Hodder and Stoughton.
Hessayon D G (1980) *The house-plant expert.* PBI Publications.
Slade J R (1981) *Notes on the keeping and using of live animals and green plants in schools and colleges,* Southern Science and Technology Forum, University of Southampton.
SAPS *(Science and Plants for Schools),* Homerton College, Hills Road, Cambridge,CB2 2PH

Animals in Scientific Research
AMRIC (1991) *Finding out about animal experiments,* (Revised edition) Hobsons, Cambridge.
BUAV (British Union for the Abolition of Vivisection), 16a Crane Grove, Islington, London, N7 8LB
NAVS 51 Harley Street, London, W1
AMRIC 12 Whitehall, London, SW1A 2DW
RDS (Research Defence Society), 58 Great Marlborough Street, London, W1V 1DD

Responsible Pet Care/Care and Maintenance of Living Organisms
Archenhold W F, Jenkins E W and Wood-Robinson C (1978) *School science laboratories,* John Murray, London.
CLEAPSS (1989) *Laboratory handbook,* CLEAPSS School Science Service, Brunel University, Uxbridge.
Kelly P J and Wray J D (1973) *The educational use of living organisms: A source book,* English Universities Press for the Schools Council "Educational Use of Living Organisms" Project.
RSPCA (1984) *Animals in schools,* RSPCA, Horsham.
RSPCA (1988) *Visiting animal schemes,* RSPCA, Horsham.
RSPCA (1989) *Small mammals in schools,* Second Edition, RSPCA, Horsham.

Slade J R (1981) See under *Plants* above

UFAW (1972) *The UFAW handbook on the care and management of laboratory animals,* Churchill/Livingstone, Edinburgh.

UFAW (1988) *The use of animals in British schools,* Second Edition, UFAW, Potter's Bar.

Wray J D et al (1974 and 1975) Publications of the Schools Council *"Educational use of Living Organisms Project",* English Universities Press. The series includes publications on Animal Accommodation, Small Mammals, Organisms for Genetics, Micro-organisms, Plants and Organisms in Habitats and a code of Recommended Practice for Schools Relating to the Use of Living Organisms and Material of Living Origin.

CLEAPSS See p.147

RSPCA (Royal Society for the Prevention of Cruelty to Animals), Causeway, Horsham, West Sussex, RH12 1HG

UFAW (Universities Federation for Animal Welfare), 6 Hamilton Close, South Mimms, Potter's Bar, Hertfordshire, EN6 3QD

Living Organisms and the Environment

Bennett D P and Humphries D A (1983) *Introduction to field biology (Second Edition),* Edward Arnold, London.

DES (1989) *Safety in outdoor education.* HMSO, London.

Dowdeswell W H (1984) *Ecology: Principles and practice,* Heinemann, Oxford.

Jenkins P F (1973) *School Grounds: Some ecological enquiries,* Heinemann, London.

Smith D (1984) *Urban ecology,* G. Allen and Unwin, London.

Williams G (1987) *Techniques and fieldwork in ecology,* Bell and Hyman, London.

The Mammal Society, Burlington House, Piccadilly, London

Nature Conservancy Council, Northminster House, Peterborough, PE1 1HU.

Dissection

Dixon A (Ed) (1988) *Issues surrounding the use of animals in science lessons: A resource pack,* ASE, Hatfield.

Lock R (1984) *A cut above,* Times Educational Supplement. January 13 1984.

RSPCA (1986) *Dissection,* RSPCA, Horsham

UFAW (1968) *Humane killing,* UFAW, Potter's Bar.

Alternatives to Dissection
    RSPCA (1984) *Alternatives to dissection*, RSPCA, Horsham.
    UFAW (1988) *Animals in science teaching*, UFAW, Potter's Bar.

Living things and the law
    DES (1989) *Animals and plants in schools: Legal aspects*, Administrative Memorandum No 1/89, DES, London
    Everett K and Jenkins E (1991) *A safety handbook for science teachers*, Fourth Edition, John Murray, London.

References

ASE (1988) *Biotechnology*, Education in Science, 126, pp12-16.

ASE/IOB/UFAW (1984) *The use of animals and plants in school science*, Education in Science, 108, pp11-12.

CLEAPSS (1989) *Laboratory handbook*, CLEAPSS School Science Service, Brunel University, Uxbridge.

DES (1989) *Animals and plants in schools: Legal aspects*, Administrative Memorandum No 1/89, DES, London.

DES (1990) *Microbiology: An HMI guide for schools and Further Education*, Second Impression, HMSO, London.

DES (1991) *Science in the National Curriculum* (1991), HMSO, London.

DOE (1988) *Protecting Britain's wildlife*, DOE, London.

Dunkerton J and Lock R (1989) *Biotechnology – A resource book for teachers*, ASE, Hatfield.

Henderson J and Knutton S (1990) *Biotechnology in schools: A handbook for teachers*, Open University Press, Buckingham.

Lock R (1989) *Investigations with animals and the Animals (Scientific Procedures) Act, 1986*, School Science Review, 71, 255, pp74-75.

Lock R (1992) *Animals in secondary school science*, Journal of Biological Education. (In press)

Lock R (1993) *Animals and the teaching of Biology/Science in secondary schools*, Journal of Biological Education, (In press).

Lock R and Millett K (1992a) *Student experience of animals outside school*, Humane Education Newsletter, (In press)

Lock R and Millett K (1992b) *Using animals in education and research – Student experience, knowledge and implications for teaching in the National Science Curriculum*, School Science Review, (In press).

Lock R and Millett K (1992c) *The Animals and Science Education Project 1990-91:* Project Report, University of Birmingham, Birmingham.

Millett K and Lock R (1992) *GCSE students' attitudes towards animal use: Some implications for Biology/Science teachers,* Journal of Biological Education, (In press).

Royal Society/Institute of Biology (1975) *Dissection in schools,* Institute of Biology, London.

Science for all:
Science for all pupils

10 Pupils with Specials Educational Needs in Mainstream Schools

Dr Bob Duerden and Alan Jury

1 Introduction

> *It is part of the teacher's professional role to recognise and develop the potential of individual pupils. All pupils should be encouraged throughout their school careers to reach out to the limit of their capabilities. This is a formidable challenge to any school since it means that the school's expectation of every pupil must relate to their individual gifts and talents.* (From "The School Curriculum" (DES 1981))

The Education Reform Act established the *right* of all pupils of compulsory school age attending a maintained school to a broad and balanced curriculum, including the National Curriculum. This right exists whether or not the child has a statement of special educational need and regardless of sex, cultural or linguistic background.

Before examining in detail how science departments might respond to this challenge, it is useful to examine the concept of "special educational needs" and to highlight the key features of the approaches to teaching and learning which are likely to succeed in meeting these needs.

2 The Nature of Special Educational Needs

The 1981 Education Act determined that a child has special educational needs if he or she has a learning difficulty which requires special educational provision to be made. A learning difficulty was said to exist if a child had:

- a significantly greater difficulty in learning than the majority of pupils of the same age; or
- a disability which either prevented or hindered him or her from making use of educational facilities of a kind generally provided in schools for pupils of the same age.

Special educational provision was defined as being additional to, or otherwise different from, the educational provision made generally for pupils of his or her age. The Act suggested that these pupils could be sub-divided into two groups. The first, larger, group would have their needs met within the normal resources available within their ordinary school. Warnock (DES 1978) suggested that approximately 20% of pupils would be likely to fall into this group at some point in their school careers. Within this group, a smaller group (about 2% of the school population) exists, whose disabilities need to be specifically addressed and for whom the authority is required to maintain a formal Statement of the child's special needs and of the special provision required. The range of special needs which may be present in such a large group is extremely diverse and it has been said that these pupils have little in common apart from a deep sense of isolation and poor self-esteem. This can arise because many pupils with learning difficulties show some of the following characteristics:

- They absorb little from conventional teaching at any one time;
- They take longer to master particular skills;
- They need to repeat new learning before they are able to retain knowledge;
- They need more practice than other pupils before they can generalise from a learning experience;
- They find it difficult to put events into a logical sequence;
- They lack the confidence to work independently;
- They have a long background of experiencing failure;
- They may develop behaviourial and/or emotional problems as a consequence of repeated failure.

The key to promoting effective learning for all, in groups containing pupils with a variety of abilities lies in presenting differentiated tasks – matching learning activities to the competencies and contexts of individuals – and ensuring equal access to those activities by removing the barriers which prevent an individual, especially one with special needs, from interacting with the learning experience. For those pupils falling into the first category defined by the 1981 act, the provision of differentiated tasks is especially important. For pupils falling into the second category the teacher must endeavour to remove any barriers preventing access to the learning experiences. Pupils do not, however, fall into neat categories and many pupils with special needs will require both a well-differentiated curriculum and assistance in gaining access to learning. It is also important to provide the right environment for learning to take place effectively. Developing such approaches will promote greater equality of opportunity and so benefit all pupils.

> *"What is good practice in relation to special educational needs is good practice for all"* (NCC 1989a)

3 Responding to the Challenge in the Science Department

Science has particular characteristics which can be of great benefit for pupils experiencing difficulties in other areas of learning. The emphasis on practical activities means that difficulty with reading and writing need not stand in the way of success, and these and other science activities based on first hand experience can capture pupils' imaginations and help to reduce behaviourial problems. Science lends itself to learning in small sequential steps, which can improve concentration and increase opportunities for experiencing success. Group working allows pupils to share in planning tasks and gives scope for improving communication skills.

Curriculum Guidance 9 (NCC 1992) states that schools should *"ensure that pupils are taught the most appropriate parts of the National Curriculum by reference to their current achievements, irrespective of their age"*. The Programmes of Study for each key stage, which should be the starting point for all planning, allow considerable opportunity for activities which are in the context of the contents for that key stage, yet can be tackled at a lower level of attainment. This enables all pupils to work on a common area of the science curriculum without being forced to experience impossible challenges. It also ensures that the least able experience changing contexts for learning rather than being subjected to the same task over and over again. So pupils can be presented with challenges which promote small steps of progression towards higher achievement.

There will, of course, also be the need to re-visit concepts at lower levels which have been encountered but not fully mastered at earlier key stages. The challenge here is to provide new, interesting and progressively more advanced contexts within which previously encountered concepts can be consolidated and extended. Furthermore, as the National Curriculum itself recognises, some of the higher level concepts specified for a particular stage may not be appropriate for some pupils. Difficult judgements concerning the science content appropriate for particular pupils cannot be avoided.

3.1 Approaches to teaching and learning

The National Curriculum for Science, and Sc1 in particular, require that pupils take an active part in the planning and execution of investigational work. This practical and pupil-centred approach can have particular benefits for pupils with special needs, with its emphasis on learning from first hand experience and the opportunity to work both collaboratively and individually on activities which build on existing knowledge and experiences. In their Non-Statutory Guidance for Science, the NCC suggest a range of roles which the teacher might assume to encourage the development of an appropriate level of independence.

Table 1 Roles assumed by teacher to encourage pupil independence

Role	Example
Enabler	Facilitates the learning opportunities which are the objectives of the lesson
Manager	Co-ordinates all class activities and organisation.
Presenter	Sets the scene, clarifies the processes involved and gives information.
Adviser	Listens, suggests alternatives, offers references and encourages.
Observer	Studies the processes and gives feedback.
Challenger	Comments critically on procedures and outcomes.
Respondent	Answers questions.
Evaluator	Assesses progress against learning objectives.

Science: Non-Statutory Guidance (NCC 1989)

This set of roles provides a setting within which a 'learning centred' approach to science education can be developed. Such an approach was suggested as part of the Children's Learning in Science Project (University of Leeds, 1987). (See also page 70) This was based on the constructivist theory of learning which suggests that children develop their own models to explain their experiences and therefore come to science with their own preconceived ideas which they use to explain the world. It is, therefore, important to provide opportunities for pupils to recognise and reflect on their models, to realise that others may have different models to explain the same phenomena, and to challenge, test and, if appropriate, modify their ideas. The teaching and learning process is seen as having 5 stages:

- *Orientation,* in which attention is focused on the area to be developed;
- *Elicitation,* which enables pupils to recognise their preconceived ideas;
- *Restructuring,* in which pupils are given the opportunity to challenge their original ideas and perhaps construct new models;
- *Application,* which offers pupils the opportunity to test their new models in a wider range of applications, to provide consolidation;
- *Review,* in which pupils reflect upon how their ideas have changed, so encouraging their active involvement in the learning process.

This approach implies that the teacher must shift the focus of attention more frequently from the whole class to individual pupils and their particular needs, and the emphasis from class-based activities to those based on the pupils' existing knowledge and experiences, in which their ideas are valued rather than being dismissed as 'incorrect'. It allows for individual differences between pupils, encourages co-operation and listening to, as well as learning from, each other, reduces the need for written materials, encourages and values verbal as well as written responses and encourages pupils to take responsibility for their own learning and to work at an appropriate pace.

3.1.1 Differentiation

A well differentiated activity will build on past achievements, offer challenges which the pupil can tackle successfully, remove barriers to learning and allow pupils to assess their own progress against identifiable criteria. A strategy which can be used to implement the constructivist approach to learning and to ensure that adequately differentiated activities are provided which promote concept development and understanding in pupils is summarised in Fig 1, because of its relevance in meeting the needs of pupils with special educational needs. This work is discussed in more detail elsewhere (Humberside County Council, 1992).

3.2 Assessment

Teachers need to develop assessment strategies which permit all pupils, including those with special educational needs, to demonstrate their capabilities and achievements. The approach to differentiation described here (fig 1)

A Model for Differentiated Science	
Select from the Programmes of Study ↓	Read through the Programme of Study for the appropriate Key Stage. Select a section on which to base the pupils' work.
Identify the Key Concept ↓	Identify the basic scientific idea underlying that section and towards which pupils will be working.
Determine the Learning Objectives ↓	List the 'sub-concepts' – the steps on the way – leading to understanding of the key concept.
Choose the Content Vehicle ↓	Decide on the subject matter to be used to extend the pupils' knowledge and understanding.
Select a Common Experience ↓	A common experience is needed to focus the pupils' attention on their common conceptual goal.
Consider Issues of Access ↓	Do any special arrangements need to be made to allow all pupils to take an active part in the learning activities?
Present the Common Experience ↓	Present the pupils with a common experience related to the key concept which will focus attention and stimulate pupil discussion.
Allow Group Discussion ↓	With appropriate teacher intervention, pupils should discuss the common experience in small groups and plan activities to test and extend their ideas.
Allow Group Investigation ↓	Groups of pupils should be allowed and helped to carry out investigations they have planned.
Encourage Group Conclusion ↓	Pupils should be encouraged to consider the results of their investigations and the implications for their original hypotheses and understanding.
Conclude Learning Activity	

Figure 1 A model for differentiated science

provides an effective means of making observational and on-going assessments of pupils' abilities without having to depend on either end-of-unit tests or whole class assessments. Good assessment also allows the teacher to diagnose areas of difficulty and so devise follow-on activities which enable pupils to overcome these hurdles. Diagnostic knowledge of the achievements and capabilities of individual pupils enables the teacher to choose suitably challenging tasks for each pupil.

Pupils will also benefit if part of the assessment process focuses on social and personal achievements. The use of interactive approaches to Records of Achievement can be particularly helpful in this respect. Good schemes for recording achievement enable pupils to recognise and value their work and to agree on targets for the next set of activities. Involving pupils in the negotiation, evaluation and review of their curriculum goals can be highly motivating. Less able pupils can be encouraged to become more involved in compiling their individual Records of Achievement by the use of a range of approaches including computers, video recording, photographs, drawings, tape recordings or collecting pictures and objects.

3.3 The Learning Environment

The constructivist approach not only implies changes to teaching style but also to the learning environment, to allow and encourage pupils to develop greater independence and take greater responsibility for their own learning. Factors to consider are:

a) The creation of an attractive and stimulating physical environment. The display of pupils' work provides invaluable opportunities for them to communicate ideas, view the work of others and reflect upon the activities they have undertaken; it is also a stimulus for further discussion.

b) The organisation of furniture within the classroom or laboratory is of great importance in encouraging pupil-centred activity. Careful consideration should be given to seating and working arrangements to encourage pupils to see the group, rather than the teacher, as the centre of activity. This can be achieved even in laboratories with fixed furniture by seating pupils on both sides of the benches, rather than all facing the front, and by the teacher not being based at the teacher's desk.

c) The effective use of increased opportunities for links with other curriculum areas.

d) The active involvement of the school SEN support staff in the science department demonstrates commitment to the provision of a broad curriculum for all pupils. Team teaching with science staff can provide opportunities for more teacher contact with pupils and opportunities for members of staff to share expertise. Involvement in planning can be invaluable in assisting with differentiation and access. Such staff can also provide in-service training for the department.

e) The development of partnerships with home, the community and industry further supports the work carried out in school. They provide opportunities for pupils to give their work a wider audience and achieve recognition from other pupils, staff, parents and the wider community. These links can also provide a greater relevance to work within school.

HMI (DES 1991) noted that the

> *increasing trend towards more use of adult support for pupils with (special educational needs) within ordinary classes...was providing some improved teaching and learning.*

Whether the adult support comes in the form of a Teacher's Aide, a Non-Teaching Assistant, a parent or other volunteer helper, it can be of considerable value in supporting the learning of pupils with special needs so long as their capabilities are known, the purpose of their support is clear and the helpers' role has been clearly planned as part of the learning process. Some schools have begun to develop support programmes using regular input from sixth form students.

4 Hints on Improving Access for Pupils with Special Needs

4.1 Pupils with Hearing Impairments

The degree of hearing impairment will directly affect the child's learning and communication skills. It is important to know which frequencies can and cannot be heard, (a chart showing this may be available: Low frequencies = vowel sounds; high frequencies = consonant sounds) and to realise that hearing aids can amplify *all* sounds.

Attention to the following points will provide hearing impaired pupils with the best opportunities to take an active part in learning activities:

- Ensure that the child's hearing aid (if worn) is functioning properly;
- Position the pupil so that he/she is facing the teacher, and encourage him/her to turn to face other pupils who are speaking;
- Gain the pupil's attention before beginning to speak;
- Remember that a hearing impaired person will probably not be able to listen and write at the same time;
- Use a communicator which can benefit the pupil greatly;
- Try rephrasing, not simply repeating, when a hearing impaired person fails to understand,
- Be aware of the limitations imposed by the room's acoustics and try to improve these (by using curtains, for example)

To achieve the greatest benefit, the pupil with hearing impairment must be enabled to participate fully in all 'concrete' experiences and allowed full access to the reasoning behind an activity and the follow-up discussion, otherwise the practical activities become experiences without meanings. Help pupils to follow discussion by referring to the equipment used and visual displays of results rather than by relying on verbal input alone.

Pupils with hearing impairment tend to have difficulties with writing, spelling, reading and language. It may be helpful to:

- Provide work that can be copied;
- Translate the pupils' dictated work to a hard copy and let them copy it;
- Use drawings and other alternative forms of recording;
- Make the maximum possible use of visual materials such as posters, photographs, diagrams, models and videos.
- Provide pictures of events for the child to sequence;
- Use 'signs' on diagrams and work sheets.
- Use IT facilities, especially word processors and concept keyboards
- Use simple, specific, straightforward language, relatively free of jargon and with key words reinforced
- Use concrete expressions when you have to convey abstract ideas.

4.2 Pupils with Visual Impairment

Identify the type of visual impairment presented by a pupil and consider the implications of this for the activities which are proposed. Red–green colour blindness may, for example, affect the pupils' ability to undertake particular chemical analyses: often pupils need to be seated close to the blackboard and/or provided with copies of notes.

Changes may need to be made to the materials presented, the equipment used, the lay out of the classroom, the teaching strategies and to the provision of support from other adults in the classroom for instance.

a) Use big, bold and brightly coloured equipment. Volumetric work may be enhanced by the use of measuring cylinders with external ridged graduations or by the use of light probes.
b) Add colour to water to make it more visible.
c) Use tactile rulers and talking thermometers
d) Find out about low vision aids.
e) With a pupil for whom large print improves reading, use a computer with a program such as *'Pen Down'*. (Would a Braille keyboard be useful?)
f) Use charts with textured materials
g) Mark demonstration diagrams using thick black felt tip pen on a yellow background.
h) Use Braille labelling.
i) Provide pupils with tactile plans of the work area (could these be made in CDT?) and help them to explore it. (Can *you* use Braille signs?)
j) Do not change the lay-out of the work area without warning the pupil.
k) Ensure the work area is well illuminated, that the pupil is seated in the best position to benefit from the lighting, and that there is not a bright light or window behind the teacher.
l) Make use of LEA specialist resources; borrow them if you can.

Additional problems which may be experienced by pupils with visual impairments are:

- Slower development of reading and writing skills;
- Difficulty in understanding concepts of symmetry;
- Perception and translational problems;
- Difficulty with spatial awareness;
- Problems with fine motor skills.

Take these factors into account when planning activities, but do not let them prevent a pupil from taking an active part, as the science lesson can be an opportunity for the pupil to develop these skills.

4.3 The physically challenged pupil
It is particularly important to enlist the help of the local support service when working with physically challenged pupils. They will not only be able

to advise on the educational implications of particular types of disability but may also have specialist equipment and aids which may be available on loan.

Have the same expectations of disabled pupils with regard to standards of effort and behaviour as you do for all. Treat the pupil who has a disability, first as an individual pupil, then, if necessary, in the context of the particular disability: the pupil is a person with a disability rather than a disabled person.

- Involve other members of the group in assisting a pupil with a physical disability – thus encouraging the development of friendships and reducing isolation: too much help should not be given, or independence may be compromised.
- The use of non-slip mats can make lesson activities more accessible to many pupils. Use might also be made of wall projection microscopes, selectagrips, scissors of various types and sizes, simpler types of connector and larger crocodile clips for electrical circuit work
- Where lifting things is a problem for the pupil, think creatively about using lighter equipment.

If some pupils with physical disabilities also have difficulties with speech:

- Remember speech therapy can often assist,
- Remember that the teacher's and other pupils' understanding of the speech will improve with time
- Consider how tape recorders, typewriters, computers, microwriters, touch-talkers and similar aids might be used

For pupils who are in wheelchairs:

- Arrange adequate space for them to manoeuvre round the room; let them sit at an appropriate, stable work surface where equipment is readily accessible.
- Consider if, for some pupils, transfer to adjustable chairs with arm supports may help (better than modifying laboratory furniture).
- Note that, if it is necessary to lift a pupil from a wheelchair it is important to be aware of her/his condition and to use correct lifting procedures, for the safety of both pupil and adult.

Pupils who have deteriorating conditions will require more assistance as they grow older. They should be given the opportunity to organise their activities for themselves as far as is practicable, and should be encouraged to direct helpers to carry out their decisions.

Where pupils have perceptual difficulties because of their restricted view of the world, (they often do), concentrate on the positive aspects of the pupil and what they can achieve, not the negative aspects of the disability. This will help reduce the tendency to allow pupils to under-achieve and encourage appropriately high expectations from others, who have contact with the pupil, pupils and adults.

4.4 Pupils who have Emotional and Behaviourial Difficulties

Emotionally disturbed pupils tend to be under-achievers with a low self-image, little self-control and poor interpersonal skills. Their experience of school tends to be negative and they are therefore inclined to be un-cooperative in the classroom. In managing these pupils, it is important to establish a positive climate in which they are encouraged to achieve and in which their work is shown to be valued. Behaviour is best managed within the context of a whole school policy which is applied consistently and sensitively by all staff.

Have a clear set of rules for behaviour in the classroom or laboratory. Although these should reflect the general rules within the school, there should be more specific ones in science: few in number, clearly and positively expressed and reinforced verbally and, when necessary, in writing. Teachers should see such rules as providing opportunities for rewarding pupils and reinforcing appropriate behaviour rather than as a means of highlighting misbehaviour.

Make particular efforts to differentiate their work and relate it to previous experience and particular interests. Present the work in small, achievable steps, which will allow the pupil opportunity for success, with consequent praise and reinforcement from the teacher. Provide further reinforcement by the frequent marking and display of work.

Where activities involve particular dangers, it may be necessary to modify the equipment or materials used. Consider using simpler, non-breakable apparatus and weaker concentrations of reagents. Where such changes are made, make them for the whole group. This avoids not only the danger of interference in other pupils' work but also the risk of isolating and alienating the pupil and so exacerbating any behaviour problem.

4.5 Pupils with Reading Difficulties

Support for pupils with reading difficulties can be provided by:

- Presenting written information in a simple, uncluttered format. Many pupils find it difficult to find their way around a typical science text-book or work sheet. They need clear directions for navigating through the mixture of text, pictures, diagrams and captions;
- Reducing text density and simplifying the language structure; using a clear cursive script with enlarged letters rather than standard typefaces;
- Using a tape recorder, adult or friend to make instructions and informa-tion available to them orally;
- Using photocopied enlargements;
- Using IT facilities with speech synthesizers, talking thermometers and balances;
- Using liquid crystal 'fever strip' thermometers rather than standard clinical or laboratory thermometers;
- Allowing work to be recorded in a variety of ways – tape recorders, cameras, drawings for example;
- Making worksheets more inviting by the use of humorous illustrations.
- Using the same title and illustrations in all versions of modified work-sheets. This enables the pupils to feel that they are still doing the same work as their peers.

5 Responding to the Needs of the Most Able

Most of what we have considered here has been concerned with meeting the special needs of those for whom learning presents difficulties. It is easy to forget that special provision also needs to be made for the most able pupils if they are to be helped to realise there full potential.

> *"The very able in schools are often insufficiently challenged"*
> (HMI 1992c)

Differentiated learning tasks again offer a solution to this problem. There should be available to high ability pupils activities which are stimulating and demanding and which will provide a foundation for further study at a more advanced level. Though such provision must, of course, include comprehen-sive coverage of the National Curriculum, high ability pupils should be extended not so much by having to accumulate an ever greater store of scientific knowledge but more by the opportunity to apply scientific principles to the real world, to investigate and solve problems and to apply scientific methodologies where the answers could not have been anticipated.

Recent studies indicate that able pupils handle the processes of science better than others. They are particularly good at deriving and controlling variables. It is therefore more appropriate to provide them with enrichment opportunities based on the application of process skills in the context of Sc1 than to extend the range of their work. The most effective vehicles for such enrichment are open-ended investigations. Such an approach will also enable them to work alongside their peer group on similar tasks but to tackle them in different ways. There may be times when it is appropriate for the enrichment opportunity to replace the core study – able pupils tend not to need to take every step along the pathway of concept development but have the facility to grasp the component parts with relative ease and so move forward in quite large jumps. Studies in the USA suggest that accelerated progress through the curriculum is effective, but this will depend on the school's policy and its ability to make such provision within the normal timetabling constraints. The key to effective provision for the most able lies in early identification of their capabilities (which may not be the same in other subject areas), the provision of suitably differentiated activities in the context of Sc1, and careful monitoring of progress. Also, the use of negotiated learning objectives and regular formative assessment and recording of achievement can be invaluable.

6 A Departmental Audit

It is useful to carry out a departmental audit to determine the fitness of provision for all pupils, including those with special educational needs. The following questions may help:

- How are pupils with special needs identified and what provision is made for them?
- What provision is there for reviewing procedures and practices in the light of knowledge about individual pupils?
- Is there effective communication between the science team and the special needs support staff/services?
- Do the curriculum and timetable permit you to deliver appropriate learning support to those who need it?

- Do current approaches promote access for all?
- Do these approaches meet the needs of individual pupils?
- Do learning outcomes reflect the teaching input?
- Have worksheets (where used) been suitably modified?
- Do book resources cater for the range of pupil ability?
- Is there clear access into and around the room?
- Is equipment accessible and logically and clearly labelled; and is there opportunity for pupils to learn the location of equipment?
- Do timetabling arrangements lead to unnecessary room changes?
- Does the basic stock of equipment need to be modified to ensure safer practical work?
- Have learning materials been reviewed to ensure that diagrams, transparencies etc. are clear and uncluttered? Are large print copies of worksheets available? Would it be possible to produce tactile versions?
- Do schemes of work ensure a clearly structured spiral of learning that incorporates the development of skills as well as knowledge?
- Do individual lesson/unit plans clearly identify the ways in which the range of abilities and special needs are to be catered for?

Dr Bob Duerden is General Adviser (Science) in Humberside, following several years as Head of Science at The Immingham School. He is closely involved with the production of guidance materials on meeting the needs of SEN pupils for both Humberside and the NCC, and is involved with the development of science facilities for SEN pupils.

Alan Jury taught in a variety of primary and secondary schools before taking up his present post at Appleton House where he is involved in providing support on behaviour related issues to pupils and schools in the Scunthorpe area. He is co-author of Humberside's guidance on Access to Science at KS3 & 4.

References

ASE (1992) *Science and special needs – a resource pack for INSET*, ASE.

Barthorpe T & Visser J (1991) *Differentiation: Your responsibility*, National Association for Remedial Education.

Croner's Guide for Heads of Science, (1988 on) Croner Publications.

DES (1978) *Special Educational Needs*. Report of the committee of enquiry into the education of children and young people. (The Warnock report), HMSO.

DES (1981) *The school curriculum*, HMSO.

DES (1991) *Science Key Stages 1 & 3*. A report by HM Inspectorate on the first year: 1989-90, HMSO.

HMI (1992a) *Science at key stages 1, 2 & 3*, a report by HMI, DES.

HMI (1992b) *Special needs and the National Curriculum:* a report by HMI, DES.

HMI (1992c) *The education of very able children in maintained schools:* a review by HMI, HMSO.

Jones A V (1983) *Science for handicapped children*, Souvenir Press.

Jones A V (1988) *Things for children to make and do*, Souvenir Press.

Jones A V (1991) *Science for pupils with special needs in the National Curriculum*, Nottingham Polytechnic.

Humberside County Council

(1990) *Science for children with special educational needs: Key stages 1 & 2*, Humberside County Council, 1990.

(1992) *Access to science for children with special educational needs: Key stages 3 & 4*, Humberside County Council.

McCall C (1983) *Classroom grouping for special needs*, National Council for Special Education..

NCC

(1989) & (1991) *Science: Non-statutory guidance*, NCC.

(1989a) Circular No.5: *Implementing the National Curriculum – Participation by pupils with special educational needs*, NCC.

(1991) *Science and pupils with special educational needs*, NCC Inset Resources.

(1991) *Science Explorations*, NCC Inset Resources.

(1991) *Curriculum Guidance No.10: Teaching science to pupils with special educational needs*, NCC.

Reid D J & Hodson D (1987) *Special needs in ordinary schools*, Cassell Education.

SSCR (1987) *Better Science: For young people with special educational needs*, SSCR Curriculum Guide No 8, ASE.

Tunnicliffe S D (1987) *Science materials for special needs,* British Journal of Special Education, Vol 14, No 2.

For readers who wish to obtain information about pupils with more severe and complex learning difficulties, the following sources are recommended:

Manchester Education Committee Teacher Fellows (1990) *Entitlement for all in practice, & Science for all,* both published by David Fulton.

NCC (1992) *Curriculum Guidance No 9: The National Curriculum and pupils with severe learning difficulties,* NCC.

Readers seeking advice on working with very able pupils are recommended to contact:

Mensa Foundation for Gifted Children, British Mensa, St John's Square, Wolverhampton, WV2 4AH Tel 0902 772771

National Association for Curriculum Enrichment & Extension, The National Centre for Able and Talented Children, Park Campus, Boughton Green Road, Northampton, NN2 7AL Tel 0604 710308

The National Association for Gifted Children, Address as No 2 Tel 0604 792300

It would also be helpful to refer to:

The Education of Very Able Children in Maintained Schools, a review by HMI, HMSO, 1992.

11 Race, Equality and Science Teaching

Pauline Hoyle

1 Why Address Equality

There are strong curricular reasons for relating science education to the issues of race and equality. The ASE Multicultural Working Party emphasised:

> *the potential of science to enhance the curriculum, by drawing on the richness and diversity of cultures and on their practice of science, now and throughout history. (ASE, 1990)*

But science is part of the core National Curriculum. If this curriculum is to be a true entitlement for all, it is essential that the curriculum is made accessible to all and is delivered in such a way as to allow all pupils to learn effectively. Racism remains a powerful force in our society and so affects pupils' learning, achievement and attitudes towards others. A science curriculum which enables all pupils to learn to the best of their ability must, essentially, address issues of equality. We must use, what the same working party describes as *"...the powerful role of science teaching for combating racism and prejudice in society."* There is a view that science is an objective, rational and value-free set of understandings and theories. However, if we look back at history, it can be seen that science has been used to justify racist practices and opinions. A powerful example of these practices is the use of the concept of 'human race'; pseudo-scientific theories which divide humanity into distinct racial groups were used to justify social and political policies in the USA, Britain, Europe and their various colonies. These theories can still be seen even today in the justification of right-wing extremist political movements. It is, therefore, essential that all science teachers and educators confront issues of justice and equality both through and in their science teaching.

In "all-white" areas and institutions, some teachers feel that it is more difficult to consider issues of race equality. It is essential that we remember that British society is, and always has been, multicultural. For many people this implies that Britain consists of groups of peoples of different ethnicities, often defined by their skin colour. However, "multicultural" does not merely refer to the "colour" or ethnicity. It is a much wider concept. Every class-room has a diverse population of pupils who bring with them a range of different "cultural understandings". Everybody experiences factors which influence their lives and understanding of the world – gender, class, ethnicity, where they live, etc. These factors contribute to defining their unique "culture".

The ASE publication *"Race Equality and Science Teaching"* defines "culture" in the following way.

> *Culture can be considered to be the social, artistic and spiritual mani-*
> *festations of personal, group, community and national identity. It may*
> *be people who consider they have a common "ethnicity" will also*
> *share aspects of a common culture, but other factors such as age,*
> *gender, class, sexual orientation, peer group, region, belief, lifestyle,*
> *job, the media etc. are as important in defining the culture(s) of indi-*
> *viduals, communities or, indeed, of society as a whole.* (ASE, 1991)

"Culture" is, therefore, an important factor in influencing the ways in which people interpret the world and hence the ways in which they learn. Science educators need to consider the influence of "culture" on the learning of all their students even in schools and educational institutions which appear on the surface to have a "monoculture".

Britain is also a "multi-ethnic" society and many schools are ethnically mixed. It is important to define what is meant by the term "ethnicity".

> *This is a relatively recent term which enables all people to define*
> *themselves by distinct geographical, cultural, linguistic and/or reli-*
> *gious origins. ......In educational terms, ethnicity can be used to*
> *define the specific needs and experiences which a child may have*
> *and which the educational system should be addressing, for*
> *example, particular manifestations of racism in a community,*
> *developing bilingualism etc.* (Ibid, p 176)

In schools where pupils and/or their parents define themselves as belonging to an "ethnic" group, science educators need to be aware that this ethnicity may influence learning. It is also important to realise that children from a whole range of ethnic groups experience racism, both overt and covert, which affects their learning and their motivation to learn.

In summary, all forms of inequality are perpetrated by people who have been through our education system. If, as pupils, they experience inequality then their ability to learn will be affected and hence their access to attainment in the curriculum. If, as people, they perpetrate inequalities, then they prevent others from achieving their full potential. All aspects of the culture and ethnicity of pupils need to be recognised, valued and used if all our young people are truly to benefit from and have equality of access to education.

2 Criteria for Ensuring Equality

Science teachers have to decide what they can do to ensure that their science curriculum addresses issues of equality. The National Curriculum offers little direct guidance on these matters. In the *Science: Non-Statutory Guidance* (NCC, 1989a) actual reference is made to the issues involved in equal opportunities although the term is not used or explained. The NCC circular No.6 (NCC, 1989b) is concerned with cross-curricular issues. It outlines three areas of consideration: cross-curricular dimensions, cross-curricular themes and cross-curricular skills. Equal opportunities is one of the cross-curricular dimensions, as is multicultural education but, at the time of writing, no further details on what is meant by these are available from the NCC.

There are three major areas which science teachers should consider in order to address issues of equality. These are

- the context in which the science is set
- the content of the science presented
- the teaching and learning approaches adopted.

2.1 The Context
The ASE lists the principles of equality and justice in the science curriculum. It states that

> *The science curriculum should emphasise:*
> - *that science is a cultural activity practised in particular social, political and economic contexts*
> - *that science is not neutral or value free; scientists are influenced by the environment in which they work*
> - *the interplay of science, technology and society*
> - *the false genetic basis by which racism is supported.*
> (ASE 1991, p 135)

One view of science that can be promoted by science education is that it is a body of knowledge and understanding which is objective and value free. A science curriculum that supports this view point might be characterised by a pupil acquiring a plethora of facts and concepts without any understanding of the context in which the science has been devised.

Another view might be one in which science has an objective understanding of the world which has been acquired by using the "processes" of science. So students are encouraged by a variety of practical experiences to discover the "laws" of science and to understand how science has been applied to social situations. The view might well suggest that any "misuse" or "abuse" of science is not the fault of the science but of the way in which it has been used by others.

A third view is one which considers science as a way of explaining the world around us by using particular processes to come to that understanding. This view would consider that scientific understandings are transient and reflect the social mores, current economic and political priorities and religious and moral convictions of society. Pupils would, therefore, be encouraged to use the processes of science but to come to conclusions about scientific understandings by considering the context from which they came.

Many science educators promote a mixture of these views in any teaching situation. It is, however, the third view which particularly supports education for equality, as suggested by the list above. This view enables students with different cultural or ethnic backgrounds to consider how scientific understandings fit within their own understandings of the world. It is, then, important for science to be presented as merely one way of explaining the world around us. All pupils bring their "cultural" understandings of the world with them to the classroom. These will include ways of explaining the things they see around them which may or may not reflect a conventional "scientific" view of the world. As can be seen from research done, for example, by CLIS, children's frameworks are very tightly held and not easily changed. If, through the science curriculum, pupils receive, overtly or covertly, the message that science is the correct view and the best way of explaining the world, then they can quite easily find that their cultural understandings are at variance with the "school", or "science" view. This is most likely to be encountered with topics such as evolution, sex education or ecology but it could also be in areas such as astronomy, where, historically, groups of peoples have explained the night sky in a variety of ways. It is important that pupils' "cultural" views are considered and used, not ignored or dismissed out of hand. This situation makes it essential that teachers actually find out about pupils' existing views. However, this needs to be done in a sympathetic and sensitive manner, otherwise pupils may feel singled out or oddly different, which could simply add to the inequalities and prejudices they already experience.

2.2 Content

One of the ways in which issues of equality can be seen to be affecting the curriculum on offer is through the content presented to the pupils. A useful list of general principles for a science education which promotes equal opportunities is made by Sue Watts (1987). This was initially developed by a group of science teachers and educators at a Commission for Racial Equality (CRE) workshop. It states that teachers should:

- incorporate a global perspective
- understand issues relating to justice and equality
- elaborate science in its social, political and economic context
- make apparent the distribution of and access to power
- make all people involved in science overt and not hidden
- incorporate a historical perspective
- start from and value the experiences and knowledge of pupils
- use flexible teaching and learning strategies and give emphasis to the learning of science
- integrate practical approaches with the work as a whole.

2.2.1 Content in practice

An example of how these criteria might be included in a current National Curriculum scheme of work could be the following:-

Key Stage 3 Programme of Study Sc4 strand (iii)

> *Pupils should investigate simple machines such as pulleys and levers and how they can be used to solve everyday problems.*

A common way of presenting this part of the Programme of Study (Fig 1) might be to take each type of simple machine such as pulleys, levers, ramps, wheels and axles, look at examples of each type and then explain how each machine works by referring to effort, load and pivot. See-saws might then be explained in terms of the law of moments. Some applications of these types of machine might also be included.

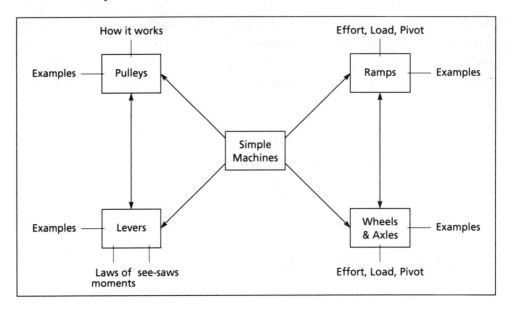

Figure 1 An approach to a "machines" topic

If, however, we applied the criteria for equality mentioned above to the same Programme of Study a different pathway would result, Fig 2. (This example is taken from some work devised by teachers for use in their National Curriculum scheme of work.)

In Fig 2, pupils own ideas about "what is a machine" are used as a starting point for the unit of work. Pupils go on to study different aspects of machines depending on the teacher assessment of the understanding shown in their original ideas. Hence the work also has in-built differentiation. To bring in the social and economic context, after this initial activity pupils might move to the activity labelled "Everyday Machines". Pupils are given some pictures/diagrams about everyday machines and are asked to consider the reasons why we use them and how they help us. Pupils may then take this further by comparing the use of machines in a house in Birmingham, England to the use in a house in Bulawuyo in Zimbabwe, (the box "Machines in Context"). This work could arise from accounts written by pupils about the machines they use in the two places they live. This introduces a realistic global perspective as well as a social, economic and political context. Pupils then apply their developing understanding by considering how a bicycle incorporates simple machines, (the box "Simple Ideas").

To introduce a historical perspective pupils could look at the use of machines in a real historical setting eg. the construction of the pyramids (the box "Machines in History"). Here it is envisaged that the story of the build-ing of the pyramids would be told in a cartoon form, emphasizing both the problems encountered in building the pyramids and the people involved. The use of different types of machines and the effect on the workforce would be considered, bringing in the issue of the use and abuse of slave labour. Pupils would consider what variety of different machines might have been used, given the resources of the area.

Another aspect of the work could be to investigate see-saws (the box "See-saw Up"). Here real situations, with pupils on see-saws trying to make them balance, would be the stimulus for work on the *law of moments*. Pupils able to develop the concept of moments in relation to the concept of work would do the activity named "See-saw Down".

The final aspect of this unit would involve pupils designing a machine for a real situation e.g. getting mangoes off a tree without bruising them. This situation would be in a real context set in Zimbabwe. It is important to use the realistic situ-ation as the resources available in the area will affect the science pupils would be able to use. It would also help pupils to realise that people are involved in every day science all over the world. Care would need to be taken to ensure that pupils did not use a negative view of the Zimbabwe situation, using terms such as "third

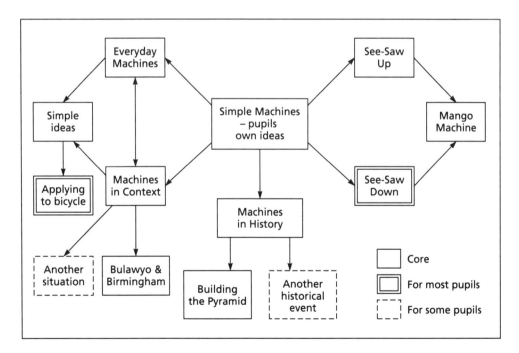

Figure 2 An alternative approach to a machines topic

world" and "underdeveloped" which are themselves negative. (ASE 1991) These are commonly held views and need redressing. If necessary, another local situation could also be used so that pupils could compare the science required in the two settings. For further information on this refer to ASE 1991 p 177.

This unit of work addresses several of the criteria for equality i.e. there is a global perspective and a social and economic context, there are people involved overtly, there is an historical perspective, it starts from pupils' own ideas. Each situation or context lends itself to meeting some criteria, but rarely all. It is important to ensure that there is a good coverage of the criteria, spread throughout a scheme of work for a key stage.

For teachers who are attempting to develop units of work of their own help in devising units is given in ASE 1991, pp 132–135. There are also some other published examples which use these criteria to which it would be useful for any science teacher to refer. See SSCR (1987) pp 17–22; Watts (1991) pp 135-141; ASE 1991, p 57;

It is not easy to find other published pupil materials that regularly address more than one of these criteria. *Science Kaleidoscope* (Hoyle et al 1990) is recommended as are some of the *Science in Process* (ILEA/HEB 1987) materials which attempt to address several of the criteria at once. The *Oxford Science Programme* (1990) certainly builds upon pupils' prior knowledge and understanding and is based upon ideas from the CLIS research. Other materials that the author has seen are less consistent and teachers would have to use them creatively in order to meet the equality criteria.

2.3 Teaching and learning

It is perhaps the methods used in the classroom that are the most influential in addressing equality. Pupils receive overt and covert messages from the way a classroom is organised and the role which they are allowed to play within that classroom. In ASE 1991 p 80 it is suggested that

> *Teaching and learning promotes equality when:*
> - *a variety of strategies is used*
> - *pupils's own experience is valued and built on*
> - *pupils are enabled to develop autonomy and responsibility for their own learning*
> - *ideas and assumptions are challenged*
> - *strategies are collaborative, not competitive*
> - *control of learning is shared between teachers and pupils.*

A classroom that works for equality will be one in which pupils and teachers are working together to enable the learning process to occur. The roles of the teacher will vary but they will enable pupils to work towards being independent and taking responsibility for their learning. Difficulties are often experienced with doing this in the secondary school, not least because of the onset of puberty, but, if it is to be done, it is the structures within the classroom which will enable it to happen.

The organisation of the classroom is important. Teachers wishing to consider equality can ask themselves questions such as

- is it a pleasant place in which to be?,
- is there a balance of pupil work and other types of displays? are the displays interactive and do they represent social, linguistic and cultural diversity?
- are cupboards labelled so pupils can have access to equipment and resources?

These will help set the tone for developing equality, but the organisation of the teaching and learning requires thought.

- who organises the seating and the way the furniture is arranged?
- do the pupils get involved in the decision making about groupings and roles within groups?
- do activities build upon pupils strengths and enable them to learn from each other?
- is there an ethos of cooperation and helping one another? are racist/sexist assumptions and behaviour challenged and discussed?
- are the resources relevant, open and accessible to all?

Teaching and learning approaches also need to be considered. In working towards equality it is essential that all pupils can interact in a variety of ways to enable them to maximize their opportunities for negotiating and developing their ideas. They need to interact with the teacher, each other, more and less experienced learners, a variety of resources and situations. This means that teachers need to develop learning environments which enhance this variety and to use a range of teaching and learning strategies over a period of time.

In the unit of work on Machines outlined above pupils had the chance to interact with each other and possibly the teacher in the first activity by talking about their own ideas. The rest of the activities used a variety of practical and text resources, each of which required pupils to interact meaningfully and collaboratively.

For many teachers attempting to apply such ideas for the first time it is difficult to envisage how to proceed. Other questions which can be asked in order to evaluate and develop a classroom designed for equality are given in ASE 1991. Some further ideas on teaching and learning are to be found in Bentley and Watts (1989) . Pupil material which addresses the issue of language development and interaction can be found in *Talking Science* (Laine et al 1989) as well as in *Science Kaleidoscope,* (Hoyle et al 1990).

3 The Way Forward

In implementing equality in science teaching there are three major levels at which teachers can address their practice: individually, in the department, and in the whole institution.

At the individual level teachers can examine their own practice in the classroom by considering the contexts and content used, teaching and learning approaches adopted and their attitudes towards the nature of science. Activities to support a teacher wishing to examine her or his current practice are presented in ASE 1991, pp 12-13, 15-17. This material deals with questions about practice in the learning environment; classroom organisation; teaching and learning; resources for learning; and expectation and achievement.

The results of such individual evaluation can then contribute, where appropriate, to discussions at the departmental level. Working together at this level can support teachers wishing to address further the issues of equality. A department can collaborate, examine their practice and develop common approaches to the following areas:

- the learning environment
- the resources used
- schemes of work; how these address contexts and content for equality, and issues of how the nature of science is presented
- the teaching and learning approaches planned in schemes of work and those actually used in the classroom
- the development of language through science teaching
- the classroom organisation
- the ethos of the classroom which encourages pupils to collaborate and share expertise
- the avoidance of stereotyping
- assessment procedures

- monitoring and attainment of individuals and groups
- rewards, sanctions and behaviour
- dealing with racist/sexist incidents

Departments can monitor their approaches, discuss them and decide which areas to tackle and improve. The formulation of an action plan which enables staff to work together on their practice, obtain appropriate INSET and support, and give time for review and evaluation is an essential part of the development of equality in science teaching.

At the school/college level equality needs to be addressed by the development and implementation of policies on the following:

- anti-racism or multiculturalism
- rewards, sanctions and behaviour
- staff recruitment
- language policy
- learning policy
- needs of bilingual pupils
- monitoring of groupings of pupils, particularly of black and bilingual pupils
- communication of the policies to staff, pupils and parents
- the mechanism for communication and sharing of policies within the school and the community.

For further details see ASE 1991, p 14

The review of these areas followed by development work will help ensure that the institution is able to offer a curriculum which addresses issues of equality and will make the work of individual science teachers more successful.

4 Summary

Equality of opportunity in science teaching will only be achieved when issues of equality are addressed in the context and content of the science presented to pupils, and when the teaching and learning approaches enable all pupils to attain their full potential. Development towards these goals can be facilitated by teachers considering their own practice, the practice and policies of their department and of the wider community of the school/college.

Pauline Hoyle is Science Inspector for the London Borough of Islington. She was a member of the ASE Multicultural Working Party and has been a Multicultural Teacher Fellow at the Institute of Education, London University. She taught science and ESL in schools before becoming an advisory teacher.

References

ASE (1990) *Race equality and science teaching,* Education in Science, January 1990, pp 67-85.

ASE (1991) *Race, equality and science teaching: An active INSET manual for teachers and educators,* ASE, Hatfield.

Bentley D and Watts D M (1989) *Learning and teaching in school science: practical alternatives,* Milton Keynes, Open University Press.

Hoyle P, Laine C and Smyth S (1990) *Science Kaleidoscope: A Student's Guide to Key Stage 3,* London: Heinemann Educational Books.

ILEA/HEB (1987) *Science in Process,* London: Inner London Education Authority and Heinemann Educational Books.

Laine C and Hoyle P (1989) *Talking Science,* Inner London Education Authority and Harcourt Brace Javanovich.

NCC (1989a) *Science: Non statutory guidance,* NCC.

NCC (1989b) *The National Curriculum and Whole Curriculum Planning: Preliminary Guidance,* Circular No 6, York, National Curriculum Council.

Oxford Science Programme (1990) Oxford University Press.

SSCR (1987) *Better science: Working for a multicultural society,* London: Heinemann Educational Books.

Watts M (1991) *Science in the National Curriculum,* London: Cassells Educational Limited.

Watts S (1987) *Approaches to curriculum development from both an anti-racist and multicultural perspective.* In: *Better Science: Working for a Multicultural Society,* London, Secondary Science Curriculum Review and Heinemann Educational Books.

12 Science for Girls and Boys

Barbara Smail

This section is an adaptation of the ASE Occasional Paper on Gender Issues in Science Education. Barbara Smail chaired the working party which produced this report. The other members were: Brenda Barber, Lynda Carr, Mary Linington, Linda Scott and Cathy Wilson (secretary), and Martin Grant and Sharan-Jeet Shan also contributed.

1 Introduction

The National Curriculum in science will ensure that all young people study some science for most of their school lives. A long term objective for many concerned with science education has been achieved. The Attainment Targets and Programmes of Study are designed to meet the needs of young people growing up in the twenty-first century. They will provide a basic understanding of scientific principles and issues for the ordinary citizen as well as for those who will work in science, technology, business and industry. At last, we have science for all...or do we?

We all know that our desire to teach must be matched by a willingness on the part of pupils to learn. Many groups of pupils have been disaffected by the science curriculum in the past. To achieve our goal, a scientifically literate population, much more than equal access to the curriculum is needed. All groups must feel that the science offered is interesting and relevant to their own lives. This section raises issues of girls' and boys' responses to the changing science curriculum. All teachers need to become aware of these differing responses, and how they are affected by the values which they are transmitting through their lessons and by the attitudes which are being cultivated in every aspect of school life.

1.1 The primary school

Although we are born with obvious physical sexual differences, we adopt our gender role mainly from unconscious pressures applied by parents and other adults during our earliest years, who may have differing expectations and beliefs about how girls and boys should be treated. When the child enters school, these expectations may be expressed as a desire to set limitations on behaviour, on access to toys, to physical activity and to areas of the curriculum, in opposition to the school's policy for equal opportunities. It is therefore important that nursery and primary teachers have discussed, as professionals, their own attitudes to equal opportunities.

It is also vital that both boys and girls have access to the whole curriculum in the primary years. Sewing, cooking, work bench, construction kits, home corner and computers should be equally available to all children. Special provision may need to be made for boys to develop home skills and for girls to develop technical and computing skills. Future interest and performance in science and technology will be affected if girls lack positive pre-school and primary experiences in these areas.

In the 1980's it was established that there were then few significant differences between the background knowledge girls and boys brought to secondary school science. But there were large differences in their attitudes and interests. These differences predisposed boys towards the physical sciences and girls towards the biological and they were enhanced rather than diminished by secondary school experience.

The continuing growth and development of science and technology in the primary school gives us an opportunity to influence the development of these attitudes at a much earlier stage. To date, there still seems to be little or no firm evidence that primary science and technology have a positive effect on interest in these areas in the secondary school. However, it is well established that in the past pupils with the most positive attitudes on entry to secondary school were most likely to opt for science later on.

In some areas of the country there has been an awareness that teachers' own attitudes to science will be transmitted to their pupils in day-to-day classroom interaction. If women and men primary teachers display differing levels of confidence and enthusiasm in teaching biological or physical science and technology themes, then early introduction of these subjects may reinforce, rather than challenge, the traditional gender bias.

1.2 The secondary school: choice and change

The introduction of "broad and balanced science" from 5-16 could produce by the year 2000 a generation of 16-year olds with less tendency to base their choice of subjects for further study and of career upon gender-related experiences. But the possibility of choice between single and double science courses sets up a situation in which there is a great danger that many more girls than boys will opt for the lighter course.

Compulsory science up to 16 may also have the effect of simply moving the age at which the gender-differentiated curriculum starts from 14 to 16. Whenever options are offered, schools must guard against building in assumptions based upon traditional gender-related patterns of choice, e.g. offering technology as an alternative to French or home economics. Computer programmes for option choices may be biased and should be used with care.

Option choices in the late 80's were still very traditional. Of girls entering examinations at 16+ in English, less than 14% entered for physics compared to 43% of boys. The percentage of girls entering for any of the technology subjects was less than 8%. There are indications that double award balanced science, delaying choice between the sciences to 16+, may be having a positive effect on the numbers of girls studying physical sciences in the 6th Form, although this is by no means certain. The *HMI Girls and Science Report* (DES 1980) suggests that the Schools Council Integrated Science Project (SCISP) had this effect, and preliminary reports from the Suffolk Science Scheme point to a similar trend.

Even when the National Curriculum is fully operational, much will depend upon the way each curriculum area is delivered in the classroom. Teachers must become aware of their own attitudes towards girls and boys and how these affect their expectations of pupil performance. For example, Spear, in Kelly (1981), maintained that science teachers display an unconscious bias in favour of boys in marking work. If this is the case, it seems unlikely that stereotyped assessments of pupils will end with the introduction of balanced science.

Choice of curriculum content can also be important. For example, it would seem to be important that in courses involving the study of materials, the classes of fabrics, plastics and metals should be given a balanced emphasis to avoid giving the impression that science and engineering are concerned only with "high technology" and resistant materials. This applies also to design and technology. For example it is too easy to find that, in project work, girls work exclusively with fabrics and plastics while boys file energetically at pieces of metal and build electronic systems.

1.3 The 16-19 curriculum and higher education

Broadening the sixth form curriculum could increase the number of girls studying physical science and technology at this stage. In Scotland, where students study 5 or 6 subjects at this stage rather than 3, a higher percentage of both boys and girls continues with physics. Currently the system of opting for a mixture of two A and two AS levels may only reinforce the arts/science divide by limiting the possibilities for future degree courses based on only 2 subjects studied in depth. The changes in 5-16 education demand adjustments in the post-16 phase and these could provide opportunities to establish greater gender equity in careers in science and technology. Those counselling students about careers and entry to further and higher education must challenge traditional role expectations. The proposed changes in 16+ education described in Chapter 16 may have a big effect.

Since the *1975 Sex Discrimination Act,* many initiatives have focused upon improving girls' access to careers in science and engineering (GIST, WISE etc) and there have been some gains over this period. However, in 1986/87, there were almost 25,000 full-time undergraduates in the UK studying engineering and technology, of whom a meagre 2,741 were women. In biological science at undergraduate level women continue to outnumber men, but they form only one-third of those studying for physical and mathematical science degrees. In computing, the number of women undergraduates actually fell from 36% in the early 80's to below 10% in 1987.

These figures continue to cause great concern among scientific and engineering employers, who are predicting severe shortages of trained personnel by the mid-1990s. The obvious solution is to increase salaries in the science and engineering sectors, but, in addition, science educators will be under pressure to increase the numbers opting for these careers, especially by tapping the pool of female talent. Ways of increasing the number of women entering these professions are now a matter of national importance. Expansion of numbers in higher education may, of course, make some difference.

2 Curriculum Issues

Equality of opportunity and access to all areas of science can be achieved only by tackling the issues of curriculum change and teacher (and pupil) awareness simultaneously. In the present climate of change, many of us are so preoccupied with other aspects of our teaching that we fail to recognise the symptoms of gender inequality which are present in our own classrooms. We also underestimate our power to redress the balance and so place the task low on our list of priorities, ensuring that it rarely gets consideration. As we continue to implement the National Curriculum, we must question constantly its effect upon girls and boys.

Teachers' ability to promote equal opportunities through their personal behaviour, teaching style and willingness to challenge discrimination at all levels varies widely. It is vitally important that all science teachers are aware that gender differences in response to science education can be changed. If science teachers work with others across the curriculum, equal opportunities strategies can be implemented successfully throughout the school.

2.1 Teaching and learning styles

In recent years, moves towards more active teaching styles, the introduction of more open-ended tasks and the greater value given to collaborative work, have at least paved the way to a learning environment which is more attractive and accessible to girls.

Placing science in a social context has been shown to enhance girls' interest. Girls studying the Salters chemistry course are more likely to continue with chemistry. Both the *Schools Council Industry Project* (SCIP) and the ASE's *Science and Technology in Society* (SATIS) (see Chapter 20) provide interesting and stimulating materials which challenge the idea that science is impersonal and abstract, unconnected with people, real life and everyday needs. Science has become less concerned with absorbing facts, passively. More emphasis is now placed on discussion, to arrive at consensus explanations of phenomena (as in the Children's Learning in Science project (CLIS)).

More collaborative programmes between schools and industry and commerce are being set up. These link schemes can provide positive role models for all young people, although care is needed in setting them up. Awareness of the possible differences in response of girls and boys to the planned activity is essential.

2.2 Textbooks and commercial packages

Authors and publishers have become more aware of the need to check texts and illustrations for balance in the numbers of girls and boys, men and women, shown. However, despite the imaginative materials being produced for integrated science, traditional textbooks for single sciences may still adopt a style which is discouraging to girls. They may portray science, particularly physics, as a world of line diagrams, without people, human purposes or humour.

Some texts have been revised by inserting token references to the kitchen while still presenting most of the science through a masculine culture of football, car engines and high speed transport. To appeal to the maximum number of pupils, books need to be produced which embrace all interests and present science in a wide variety of contexts.

Schemes of work produced by departments and individual teachers may suffer from the same shortcomings as commercially produced material. It is important that science departments discuss the issues involved so that everyone is aware of the potential pitfalls before they start work.

More materials showing women using and developing science at all levels are needed. Since few women had the opportunity to study science and technology until the latter half of the twentieth century, men predominate in the historical examples of scientific discovery and technological innovation shown in textbooks. These concentrate on "great" scientists and tend to ignore the use made of the discoveries by ordinary people. For example why not mention the nurses who used Lister's carbolic acid as a disinfectant as well as Lister himself. When discussing scientific discoveries, their social context needs to be explored and developed to explain the predominance of men in science at the time. The hidden contribution of women scientists in history, described in Alic (1986), provides an insight into the attitudes of the past towards such women and will suggest starting points for discussion.

2.3 Assessment

In 1988 the National Curriculum Science Working Group advised that care should be taken to avoid ethnic and cultural bias in interpreting test results. Gender-related differences were not specifically mentioned in this context, but girls and boys can be viewed as products of feminine and masculine subcultures. The APU report *"Girls and Physics"* (1986) suggests links between the subcultures created by the domestic and social habits of girls and boys and their relative strengths displayed on certain assessment tasks. The TGAT Report (1988) highlights research reports that girls tend to perform better than boys on open-ended tasks requiring extended writing, while boys perform better on multiple choice questions. However, it has been argued that the question of format (whether multiple-choice or free response) is less important than the content and context of the question in creating real gender differences in response scores. Girls and boys may interpret questions in different ways. For example, when young children are asked to design a boat to travel round the world, girls focus on the living arrangements aboard the boat, while boys concentrate on the external design of the boat, its engines, keel etc. These equally valid answers to the same question may be valued differently by the examiner, particularly if she/he expects a particular answer in the context of a science and technology test. (It may be of course that a different wording of the task would help.)

Teachers' attitudes to girls' and boys' work have been studied by Spear (in Kelly 1987). She found that work attributed to boys was rated more highly by teachers than identical work attributed to girls. Both male and female teachers marked work thought to be by boys more highly for "richness of ideas". "scientific accuracy", "organisation of ideas" and "conciseness". The only characteristic on which work supposedly from girls was rated more highly by the teachers was "neatness".

Against this background, we can only advise that test construction should be approached with extreme caution. Test items should be trialled, and those producing large gender differences should be examined carefully to distinguish the underlying cause. A variety of question formats and contexts should be included whenever possible.

2.4 Classroom environment and atmosphere
The location of a laboratory or classroom and its fixtures and fittings will influence the range of activities which can be carried out in it. The style in which the room has been personalised, for instance by the use of bright colour schemes, plants, posters and displays of pupils' work, will also affect its atmosphere. Is the atmosphere warm and welcoming and conducive to enthusiastic imaginative work?

A number of research studies have pointed to girls being "crowded out" by boys in many ways in science lessons. Boys may occupy greater areas of the work space or they may monopolize equipment. In either case, different strategies for allocating places and equipment are required.

Within the classroom, teachers must endeavour to pay as much attention to girls as to boys, and encourage boys to listen and girls to contribute, and vice versa. There is ample evidence from classroom research that we, as teachers, spend more of our time coping with the noisy boys; even with conscious effort on our part, we still tend to give boys more of our attention. When teacher appraisal becomes more widespread, this aspect of teaching needs to be included in the appraisal system so that we will all become more conscious of the biasses we show within the classroom.

The availability of materials, access to balances, computers and other equipment within the classroom, workshop or laboratory needs to be monitored in order that equality of experience is ensured. Teachers must also be vigilant in their attitudes towards girls and the use of tools. Girls need experience in handling tools which they do not use at home. They do not need someone else to do the job for them! Teaching strategies which boost girls confidence must be developed. If their lack of experience is treated insensitively, their insecurity will be reinforced.

For group work, pupils should be given the opportunity to work in a variety of groupings; single sex, mixed by ability and mixed ability. Both boys and girls must have the chance to lead and to follow. Teachers and pupils need to develop respect for one another. Pupils will emulate the attitudes shown by members of staff. If an ethos of equality pervades teacher/teacher and pupil/teacher interactions in the school, then the female pupil/male pupil interactions will reflect this equal valuing.

All of these factors can be manipulated by teachers to ensure that both girls and boys feel equally comfortable with the science and technology curriculum and the way it is transmitted and assessed in school. Schools can and do make a difference.

Checklist for Good Practice in Laboratory and Classroom

- Recognise that gender bias obstructs the attainment of educational objectives.
- Recognise that low involvement of girls in science and technology is a problem, but one that can be resolved.
- Be aware of pupils' earlier experiences (or lack of experience) in the subject and devise strategies to boost confidence and remedy gaps in pupils' knowledge and experience.
- Have the same expectations of girls and boys. Do not allow the sex of the pupil to influence your assessment of their potential ability.
- Be sensitive to the problems faced by a small group of girls in a class made up predominantly of boys, or vice versa.
- Be aware of the amount of time spent with girls and with boys. It may be necessary to reorganise the pupils' seating arrangements or modify teaching stance and presentation.
- Draw on the common interests of girls and boys when presenting subject matter, and avoid sex-stereotyped assumptions.
- Discard sexist teaching materials.
- Check assessment instruments for sex-differentiated outcomes.
- Monitor outcomes against objectives.

2.5 Career expectations and role-modelling

Pupils's early ideas about careers and their future lives are often based upon unrealistic assessments of their own abilities. In the GIST project, (GIST 1983), boys tended to over-estimate how good they were at science and technology, while girls under-estimated their performance. Faced with option choice at 14, many girls closed doors which could not be reopened very easily later. With a common curriculum to 16, the increased maturity of pupils at the point of option choice should mean that the courses selected reflect a clearer understanding of personal strengths and weaknesses.

Pupils' thinking about careers is influenced by many factors, including their parents and the media. In the primary school, children should be presented with printed materials, films and videos showing people in jobs and roles which are not traditional for their gender. Whenever possible, these materials should be supported by visitors who reinforce the idea that women can be engineers, doctors, bus drivers, car mechanics etc, while men can be secretaries, nurses and look after young children.

In secondary school, the present emphasis on work experience and schools–industry links should mean that pupils gain a more up-to-date picture of the workplace. However, there is a danger that some of these experiences may not challenge traditional job segregation and the tendency for women to opt for subsidiary roles. By observing current practice, young women may get the message that in industry they will be confined to support service areas. They may see women only as secretaries, personnel assistants, librarians and junior laboratory staff rather than as professional scientists, engineers and managers. In setting up work experience, teachers and their industrial partners should ensure that ideas of "suitable" jobs for women are challenged. Whenever possible, women with high status technical and managerial posts in companies should be on hand to act as role models for the girls and to modify boys' traditional attitudes. Women who have taken career breaks or who make use of work-place nurseries or job sharing are particularly useful as role models.

Teachers also act as role models for the pupils with whom they interact daily. The predominance of men in the more powerful posts within schools has a clear message. This pattern of men 'in charge' is obvious even in primary schools where, in 1985 more than 75% of women teachers were on scales 1 and 2 (compared to 36% of men) and over 50% of men teachers were heads or deputies. In secondary schools, where female and male teachers are more equal in numbers, 69% of women and 42% of men were on scales 1 and 2. At that time, 30% of men teaching in secondary schools were on scale 4 or above, compared to only 11% of women teachers.

3 Implementing Policies

The Association for Science Education in its policy statement on Gender and Science Education (ASE,1992) states that

> *"All groups concerned with science education should have explicit written equal opportunities policies to guide, monitor and evaluate practice".*

The following checklists were developed during the process of producing this policy statement. They spell out the detailed implications in terms of actions to be taken.

Checklist for teachers.

The ASE recommends that science teachers... adopt teaching/learning styles which:

- start from and build upon the personal experiences of pupils;
- enable pupils to test out their own ideas;
- enable pupils to practise a wide range of skills in a non-threatening environment;
- provide opportunities for working in groups whose constitution may vary, as well as for individual work;
- provide opportunities for pupils to use expressive language;
- provide opportunities for pupils to make evaluative statements about the role of science and design and technology in society;
- develop the "knowledge and understanding" aspects of the course in the context of society and personal responsibility and morality;
- involve pupils in discussion about sex-stereotyping, and the desirability of change;
- emphasise that achievement does not depend on gender;
- provide role models through the use of written materials relating to the lives of contemporary women scientists and technologists.
- also bring women with these careers into the classroom, laboratory or workshop.

- reinforce the importance of science and design and technology education for girls and boys of all aptitudes and abilities.
- emphasise the importance of science and design and technology qualifications for a wide range of careers.
- evaluate continually their personal practice in the light of discussion with colleagues and the production of the department's policy (see below) to identify possible sources of discrimination.

2. Checklist for science departments
The ASE recommends that science departments:

- negotiate a written equal opportunities policy which includes an action plan drawn up in the light of a departmental audit, a schedule for its implementation and a mechanism for evaluation of its effectiveness; (Such a policy and audit should relate to all within the department, pupils/students, teaching staff and ancillary staff.)
- review their curriculum schemes to ensure that they emphasise that science and design and technology are human activities which relate to people's lives;
- review their teaching schemes to ensure that they compensate for any lack of early practical experiences;
- review their resources to ensure that stereotypical views relating to adult roles and to aptitude and ability in science are not reinforced;
- consider the values that underpin their assessment schemes, giving greater emphasis to diagnostic assessment that enables youngsters to discover their own strengths and weaknesses;
- review their working environments to make them aesthetically more appealing through the display of materials, etc, in particular pupils' own work.

3. Checklist for whole school policy
The ASE recommends that with regard to pupils/students, schools:

- develop a written equal opportunities policy statement which is supported by science teachers and put into practice in the science teaching in the school;
- construct frameworks for choice, whilst option systems exist, which do not reinforce sex stereotyping;

- ensure that teachers of science and design and technology, including primary school teachers, are provided with INSET/professional development opportunities to increase awareness of gender issues;
- ensure that within teacher appraisal schedules, there are questions relating to the implementation of equal opportunities policies in the classroom, laboratory and workshop;
- provide parents and governors with information about the need to increase the achievement and participation rates of girls in science and design and technology;
- provide parents and governors with information about strategies being adopted by the school to ensure that girls participate fully in science and design and technology activities;
- provide girls and boys with opportunities to interact with people working in non-traditional employment areas with a science base.

4. The ASE recommends that, with regard to teachers and ancillary staff, schools should:

- develop a written equal opportunities policy statement which is reflected in appointments for all departments within the school;
- prepare job descriptions setting out the duties and responsibilities of all posts, whether teaching or ancillary;
- advertise all vacancies externally and internally at the earliest opportunity

Barbara Smail is Education Manager of the British Association for the Advancement of Science. She has worked in industry and taught chemistry. She has carried out research into gender differences in science education and chaired the ASE's working party on the subject.

References and Reading

Alic M (1986) *Hypatia's heritage: A history of women in science from antiquity to the late nineteenth century,* The Women's Press.

APU (1986) *Girls and physics report,* S Johnson and P Murphy: Occasional paper No. 4 (Available from the Centre for Educational Studies, King's College (KQC), University of London, 552 King's Road, London SW10 0UA).

ASE (1992) ASE Policy: *Present and future – Gender and science education*, ASE.

ASE (1990) *Gender issues in science education*, ASE.

DES (1980) *Girls and science* (HMI Matters for Discussion series, No. 13), HMSO.

GIST (1979-83) *Girls into Science and Technology project*, Manchester Polytechnic/University of Manchester.

Kelly A. (1981) *The missing half*, Manchester University Press.

Kelly A (1987) *Science for girls?*, Open University Press.

Murphy P (1986) *Differences between girls and boys in the APU science results*, Primary Science Review No. 2.

Myers K (1987) *Genderwatch! Self-assessment schedules for use in schools*, SCDC Publications.

Smail B (1984) *Girl-friendly science: Avoiding sex-bias in the curriculum*, Schools Council/Longmans Pamphlet Series.

TGAT (1988) *National Curriculum. Task Group on Assessment and Testing: A report*, Department of Education and Science and the Welsh Office.

WISE. Women into Science and Engineering. Campaign organised by the Equal Opportunities Commission and the Engineering Council.

Science for all:
Science for all ages

*In the chapters that follow we are addressing the issue particularly of conti-
nuity and progression. The authors of chapter 13, 14 and 16 have been asked
to give a flavour of what science is like at their levels, KS 1 & 2, KS 3 & 4,
and ages 16–19 respectively, in a way that seems appropriate for that level.
Chapters 13, 14 and 15 are also in the ASE Primary Science Teacher's
Handbook, so that they provide common ground to primary and secondary
teachers.*

13 Key Stages One and Two

Alison Bishop and Richard C. Simpson

In this chapter our main concern is science in the primary school at Key
Stages 1 and 2. However, because the basis for much future development
happens in nurseries, our approach is to focus on science in the nursery
and the early years' classrooms. We will then examine development, progression and continuity to KS2. This approach also reflects our view that
experiences in science in the earliest years are vital for girls as well as
boys: the ideas and attitudes which are formed here will have a profound
influence on the decisions girls make with regard to science in later
years.

We offer a series of illustrative science encounters for children which we
have found to be effective learning strategies, with some comments on the
way children learn science.

We are convinced that children of any age are perfectly capable of practising the skills identified in Sc1, provided the task is set in familiar terms.
Activities at KS1 and 2 and in the nursery, involve practical experiences
within the context of class, group and individual work, and often involve
adults from the local community. But the skills of investigation must be
linked closely to an understanding of the child's initial ideas, and any
change or development of these ideas must make sense to the child.
Investigation for the very young child must be grounded in the exploratory
play situation where the child's own individual experiences are stressed
within the task.

Practical activities provide a lot of motivation; another strong motivator is story. It is worthwhile, especially at KS 2, to tell stories of the lives of scientists and their discoveries, which include women scientists and those from other cultural traditions.

1 Water

1.1 Water in the Nursery

An area such as water play must ensure that the child is being offered a progressive series of problem solving experiences, which can be carefully monitored and planned for the individual child. A starting point could be a simple story such as the Aesop's Fable where the bird wishes to obtain a drink of water from a jug but finds the water level too low. This simple story offers to the very young child the opportunity to use materials which are simple, safe and familiar. Young children enjoy the feel and touch of the water upon their skin and will observe carefully and predict outcomes for the welfare of the bird.

Start with a series of containers of the same size and with equal amounts of water. Even the youngest child will soon shout if s/he does not consider the same amount of water is present in each container. The children can then find objects which will make the water level rise. Although some authors acknowledge the egocentricity of the nursery child, we have found that this sort of activity leads to co-operation. The opportunity must be given for children to work together in order to promote and develop their language skills: an adult can increase the complexity, quality and quantity of this talk. We have found children will remain interested and committed to this task for up to one hour. Photographs of the children working at their activities in the water trough would be useful for assessment and future planning.

Questions

It is essential that adults know when to intervene, have thought about the questions they would ask and have an idea of the kind of questions children are likely to ask. We acknowledge also the need to be flexible, and realise that children themselves may ask unanticipated, complex and mind-stretching questions. Do not underestimate the ability of the three year old. In one nursery, a girl came up to the group who were trying to raise the water level in a container with some corks, and commented, *'If you go outside and get a big stone from the yard and bung it in there, the water will soon come over the top'*.

The following questions have been tried and tested in the nurseries of South Tyneside. The children were putting objects into containers of water:

- Which objects make the water level rise the most?
- If you put more things in, does the water rise more?
- What happens to the water level if you force the polystyrene down and:
  (a) keep your hand there?
  (b) let go?
- Place a measure of pieces, such as a cup full of wooden bricks, in one container. Place a measure of polystyrene pieces in the other container (with equality of container and water). Is there a difference in the level of the water?
- If we put the same amount of water in different sized containers and 4 bricks in each, does the water rise to the same level?
- If you have 3 containers – different sizes with equal amounts of water and the same number of objects – what happens? Which things make the water rise?
  (a) things that float?
  (b) things that sink?
- Which things sink, which float?

Activities such as these, coupled with the strong curiosity shown by children, will obviously lead to a variety of experiences which lay down much of the foundation work for study in the area of materials and forces in later years.

The concepts explored in this work would include:

- Some things float, others sink.
- The position of floating objects in relation to the surface of the water varies.
- Shape and mass have a distinct bearing on this.
- Objects which sink displace water and raise the water level.

1.2 Key Stage 1

Further experiences with water should be provided throughout the infant school and should not cease after the reception class.

Another exploration of how water behaves could use tubes and funnels, sieves and other containers with holes and leaks which show children how water moves and flows and how, if put under pressure, it can move quickly and exert force. Issues such as whether the length of the tube makes the water travel further, and how the direction of flow and speed of flow can be altered, can be discussed.

Questions to initiate discussion might be:

- What happens to the water when you stop pouring?
- Why does the water move when you start to pour again?
- What happens when you raise the height from which you pour?
- If pipes are coiled at the same level does restriction of flow occur?
- How can you stop the water from coming out of the tube?
- Is anything else moving along the tube with the water? (air bubbles?)
- If there is a bend in the tube; does it restrict the flow of water?
- If you pour water in, what happens to the water in the bend?
- When wide and narrow tubes of the same length have water poured into them via a funnel starting at exactly the same pouring time and height, which water comes through first?

The concepts introduced include:

- water is a liquid
- water usually flows in a downward direction
- water finds its own level
- water can break up into several flows
- water can push on objects i.e. exert a force to hold things up.

These activities work towards the NC requirements in the following ways:

- Sc3 Materials and their properties:
  PoS i, exploring everyday materials
  SoA 3/1a, be able to describe the simple properties of familiar materials.
- Sc4 Physical processes:
  PoS iii, They should explore floating and sinking...
  SoA 4/1b, understand that things can be moved by pushing or pulling them
  2c, understand that pushes and pulls can make things start moving, speed up, slow down or stop.

1.3 Key Stage 2

A logical progression from these early enquiries would be to develop ideas from 'crossing water'. Work could lead into boats with such experiences as:

- What material makes the best boat?
- Does the shape of the boat make any difference to the way it moves ?

- How much can a boat carry?
- What happens if we load the boat on one side?
- Can we stop the boats from 'falling over'? Here, the ideas of a keel might be introduced, which produces a force whose effect is to turn the boat back upright.

Other *'crossing water'* explorations might develop from bridges and aqueducts. The poem *'The Owl and the Pussycat'* might be a good starting point to promote children's ideas about boats.

For example:

- Sc3/3a, be able to link the use of common materials to their simple properties.
- Sc4/3c, understand that forces can affect the position, movement and shape of an object.

2 Living Things

It is essential in the early years that activities include a balance of both physical and biological science. Sensitivity to living things is linked closely to their care. In the case of young children, this needs to be taught. The authors remember vividly the case of the three year old, who, after watching her first ladybird in amazement and pleasure, tried to pick it up between her thumb and finger, only to watch in horror as it was squashed. Her cry of 'put it back together again' brought words of sympathy from the other children. As well as the obvious distress caused to the ladybird, children need to be taught that it can be dangerous to pick up creatures, leaves or other living things and that tools should be used at all times.

Observation
In the earliest stages of learning, observation and communication about the living things which children find in their environment are crucial. Observation skills in particular do not come easily to the young child and s/he needs to be directed to lots of practical activities in order to gain such expertise. These activities, of course, must enable children to use all of their senses, for it is only then that young children will be able to hypothesise and test out their observations. As knowledge and concepts grow, they will influence what the child observes. They should be encouraged to observe the similarities between living things and look for patterns, as well

as noting the differences between things. Children who have the evidence of their own senses will raise questions and check their ideas against the behaviour of living things. However, they need time to discuss these ideas with a listening, attentive adult and also with other children. Explaining their ideas to others helps in the planning as well as the refining process, and we have found that the very young children, even in nursery, will check their findings, review their activities and then develop and extend their ideas. The ethos of the classroom is crucial to this development. Some of the best work seen with regard to the study of living things takes place in inner city schools where teachers have limited resources and materials and yet provide worthwhile experiences for children to interact with living things.

3 Life and Living Processes

Activities can arise from various sources but Sc2, "Life and living processes", tends to reflect traditional biological thinking with particular reference to flowering plants and human life processes. One important aspect of any biological understanding is the concept

- Structure of an organism is related to function.

The following approach is somewhat different from the traditional one of considering the organs and systems of flowering plants, and people and their functions.

3.1 Nursery and KS1

A trip to the shops is always a useful starting point. Here the children can buy both traditional and exotic fruits and vegetables. Buy tomatoes, apples, pomegranates, peppers as well as pineapples and star fruit. The children will need very little encouragement to handle, smell and taste the fruit and some interesting descriptions will follow. Printing with sections of the fruits and vegetables is enjoyable and leads to interesting discussions about similarities as well as differences. It is particularly useful, after the children have examined the fruits and vegetables for taste, colour and texture, to look inside for seeds. The concept that plant life begins with seeds is a useful one for the children to acquire.

Some of the seeds mentioned above however tend to take rather a long time to germinate, which can be very frustrating for young children, who tend to dig them up to see what is happening if growth doesn't appear quite rapidly. It's a better idea to use some quick growing seeds such as mung beans. These can be bought in quantity from Health Food Shops. Hopefully, the children will discover that plants need certain conditions before they will grow.

In order to extend their ideas to animals (i.e. any organism that is not a plant) it is useful to watch the life cycle of the blowfly. Gentles (maggots) can be purchased as bait from fishing tackle shops. These need only be kept for a few days to observe the pupa (chrysalis) then the emerging adults. So, apart from the egg stage, they can see easily, and with no problem of feeding, an animal life cycle.

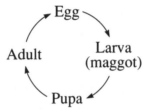

The concept that animals begin from eggs can be further explored with the eggs of butterfly, chick or frog spawn. (See ASE *"Be Safe"* on using living organisms).

3.2 Key Stage 2

Again, considering the links between the structure and function of organisms, an interesting idea is to start with a question such as, 'How do seeds get spread?' We took 8 year old children out for a seed collecting walk in the autumn. On returning to the classroom, the children sorted the seeds into those that were brightly coloured (haws, rose hips), those that might 'fly' and those in pods. We guessed that animals might eat the seeds in pods and fruits, and established the connection that much of our food comes from seeds. We looked at wheat grains and breakfast cereals. An interesting question arose here as to whether little seeds which contain comparatively less food, grow quicker than large seeds, which contain lots of food. We next talked about dispersal; winged seeds such as sycamore and ash make terrific experimental material. We challenged the children to make a big model with card and plasticine which simulates the behaviour of the real seed, but not before they had been encouraged to look carefully and compare the seeds. It is also useful with regard to observation skills to ask the children to draw what they see. From our experience, these questions may arise:

- Why is the leading edge of the wing in both ash and sycamore thickened?
- Cut it off to see the effect this gives or alternatively make the model without a thickened edge. Compare.
- Why is the seed at one end?

Make a model with varying amounts of plasticine on the end.

- Do each of the seeds perform in different ways when dropped?
    Extend the idea by investigating dandelion parachutes.
    Do these really behave like the made structure?
- How do seeds get into the ground after they have fallen? Try putting a drop of water on a wild oat seed. It will twist around and bury itself .

Concepts:

- Seeds need to 'flow' away from the parent plant
- Plants may need help to disperse seeds
- Animals may be attracted to seeds by their colour.

3.3 The Environment

We would further wish to stress, however, that any activities of life and living processes should be linked with environmental issues. Young children are interested in what is happening to the world around them. It is especially interesting to note that children who have undergone abnormally upsetting emotional and social difficulties in their own lives can develop respect for other living things, for their surroundings and ultimately, for themselves.

A recent experience in a nursery class is characteristic of the way young children react. Specimens of shells and seaweed had been collected (together with a supply of seawater) to introduce the children to the organisms of the seashore. Shells are a particularly good starter activity since no damage is done to the environment by collections of empty shells which can be used to develop observation skills. A group of children was organised to look at the plants and the shells. Green seaweed was placed in seawater. Forget the shells! A tiny shrimp darted out of the weed and the children got very excited, playing gently with it, watching closely what it did and greatly extending their use of language. They made some verbal comparisons with caterpillars which they had just studied (both are *arthropods* – jointed limbed-animals). More and more children gathered round attracted by the cries of excitement. It became virtually a class experience. One boy, who rarely settled for long at any activity, remained fascinated for a whole hour, a reinforcement of our previous experiences of the ability of young children to maintain concentration for long periods of time. It remains true that 'chance favours the prepared (open mind)' and much environmental work involves chance encounters with animals and plants.

Recent studies have shown that frequently children assess "animal" as meaning "creature with fur" so it is necessary to dissuade children of this by wide experiential work. "Animal" is simply anything not a plant in the scheme of nature. Fewer children are confused by the term plant, but again, they need experiences of other kinds of plants, eg. those that do not flower.

4 Summary

We have tried to give a flavour of science in primary schools and stress progression, investigative work and concepts from nursery through to the end of KS2 We would reiterate our commitment to play, from the nursery throughout the early years of schooling, but would acknowledge that the children can be in a situation in the classroom where they are active without being purposefully employed in the learning of science. We have therefore attempted to link the play of the child to the enabling role of the teacher, to ensure the careful planning necessary to address the concepts, attitudes and skills which are to be developed. Finally, we would repeat our concerns that, from nursery through each of the key stages, the expectations and experiences which are offered should be equally suitable for both boys and girls.

Alison Bishop is a Senior Lecturer in Education at the University of Northumbria where she teaches Early Childhood Studies in Science after twenty years' experience in teaching science to young children.

Richard Simpson is Senior Lecturer in Science and Environmental Education at the University of Northumbria. He was formerly an advisory teacher for primary science and has experience of teaching science and technology in primary, first and nursery schools.

Reference

Science in the National Curriculum (1991) DES and WO.

14 Key Stages Three and Four

Terry Parkin and Jenny Versey

The aim of this chapter is to give a flavour of, and a feel for, science education in the secondary school. We approach this task by giving an account of what one day in the life of a science teacher, a head of faculty, might be like. This is not a real day, but the account draws on our collective experience of schools and lessons in such a way that a series of important issues in secondary school learning and teaching is addressed. In taking this approach, we recognise that it is difficult to put one's finger on a typical school or a typical lesson. Enormous variety exists and it would give a false impression to assume that all experiences are the same.

What we say reflects our view that much of what is considered to be good science at secondary level is very similar to good science at primary level. With the advent of the National Curriculum and Local Management of Schools, primary and secondary teachers have become more aware of common approaches and aims, and we feel that a greater sense of urgency over liaison between phases of schooling is developing, something which we welcome. We hope this case study may help to further this process.

A Day in the Life of One Science Teacher (Head of Faculty)

Period 1: Non Contact

As I return from seeing the technicians, I look through the door of a laboratory. Something is wrong: there should be a Year 8 class here, but the room is silent and appears empty. The room smells of cloves. I soon realise that my first hypothesis, vampires, is wrong: it is cloves, not cloves of garlic; I am a physicist and I still cannot get this biology right. "Toothache" I think: perhaps someone has been using the wrong brand of toothpaste.

The class turns out to be in the corner, huddled around the radiator, gently rocking from side to side. The sun is out; the room is warm; probably not then a post LMS ploy to save money. A teacher's voice comes from the centre of the scrum.

"And what happens if we get more energy?"

The huddle begins to rock more violently. Robert detaches himself from the scrum.

"I've evaporated", he says, grinning at me over his shoulder. (Robert *would* be the first to evaporate I think to myself). He sits on the side bench.

"It's the coldest part of the room. I'm condensing on the window sir." I nod sagaciously.

"And if we get more energy?", my colleague asks the group.

The huddle starts to break up, with pupils moving around the room in organised chaos. They begin to condense back onto their stools, while the teacher remains in the corner with a watch glass in her hand. Aha! Oil of cloves, particles spreading like pupils!

The thoughts in the minds of many pupils seem to appear bubble-like over their heads. "So that's what she means when she talks about particles and energy".

However a few bubbles are clearly saying "I really enjoyed running around the room for a couple of minutes!".

My colleague sees me for the first time.

"Hello; year eight particles" she says, smiling, feeling that some sort of explanation is needed.

Several different work sheets appear to be in use. The "running round the room is fun" group have a table to complete. I see three headings, "Gas", "Liquid" and "Solid" down one side, with blank spaces opposite them and key words at the bottom of the page for them to match to the appropriate state.

A second work sheet has drawings of particles in various materials and supports a second matching exercise, drawn from a different science project, designed to consolidate understanding. Robert's group appears to have moved beyond these activities. They are attempting a range of problems on changes of state.

"Of course you can get liquid carbon dioxide" says one; "It's mixed with drinks to make them fizzy."

The teacher sits with this group. She has the gleam in her eye of a committed reconstructivist, and starts questioning them. A recent in-service course reminded her of some of the strategies she might use in modifying the concept models held by pupils.

Now that I am in the room, I take the opportunity to work with one of the groups. As Head of Faculty, I get twice as many non-contact lessons as most of my team of eight. We operate an "open-door" policy and regularly drop in to each other's lessons. I drop more than most. I read some sentences to Stuart and Jane but their understanding fails. I change some of the words and use a clever (?) example.

"How do Eskimos make soup?"

"They would get some ice and melt it, then add fish or something and boil it up."

"What would happen if their soup kept boiling?"

"It would boil away."

"So where would the water go?"

"All around the igloo. Ah."

We work out that the water is then a gas, and in a gas the particles move very fast and have lots of energy (good). There is a big distance between the molecules (excellent). I think Stuart and Jane have the pattern. They record their success in their personal record books and feel ready to attack another problem and select the next work sheet for this theory lesson.

This is a fairly typical year 8 lesson. They are mostly held in purpose-built laboratories, but, because of timetabling constraints, a group occasionally finds itself in a classroom for one or two of its four fifty-minute lessons a week. Our schemes of work usually contain a scattering of theory lessons that do not really require a laboratory. This particular group gets all four of its science lessons (as two doubles) in specialist rooms. (Examination groups are always in laboratories.) The scheme of work for particles is a component of our "Energy" module and is based on a nationally available scheme. It is closely linked to the Y7 courses of our eight feeder schools and is co-ordinated across the Key Stage by the KS3 co-ordinator, who receives an A allowance. Our schemes are very detailed and aim to support faculty colleagues in teaching unfamiliar areas of science. They include lesson objectives, (based on our interpretation of the National Curriculum) suggested means of delivery, resources, practical ideas, homework, and provision for special needs and other forms of differentiation, including work for the more able as well as the slower pupils. Once, after struggling to set "cover" work when feeling like death (well I felt pretty rough anyway) I suggested including cover work in the schemes; it may have been my most popular idea ever.

Other high schools in our area have similar cross-phase systems: in one, the Head of Science fulfils the role of co-ordinator, as she believes that the quality of the cross-phase links determines the success of her KS3 results. The LEA has provided considerable support in building up these links, but priorities are moving elsewhere. This "seed corn" expenditure has left us in a good position to develop them further, but how does this fit with the other priorities of the faculty?

Local primary schools may feed five secondaries; with coordinators in a range of subjects wanting to meet individual class teachers, we continually sympathise with the workload of primary colleagues, (even if we occasionally act as if we are the only school in the area that they send pupils to).

As the lesson comes to an end, several pupils refer to their personalised record books and note their progress for the session. Several record that they now understand particular aspects of the course, others that they need to revisit ideas, rating their progress on a 0–3 scale. These aspects started off life as Statements of Attainment but we found them unhelpful. Instead, working with support staff from our special needs department, we have developed our own pupil-friendly outcomes to guide progress and to aid record keeping.

Towards the end of this half term, pupils will be expected to produce a profile in cooperation with their teacher. (Fig 1) The profile is expected to be a positive record of achievements and we encourage all pupils to set targets for the next stage of the learning process. For about half the class, targets will be negotiated with the teacher on this occasion; the remainder will negotiate on their next profile. Typically, one profile is produced each half year, as a minimum. Two copies of the profile are raised, one for pupils, which they keep with their personal portfolio, and one for parents. The portfolio forms part of the agenda for our subject review evenings to which both parents and their children are invited: we have found that this arrangement makes these meetings most constructive.

It is ten minutes until the end of the lesson. I thank my colleague for "allowing" me in her lesson and head for the faculty office. I pass one of our two technicians. She reminds me that two orders are on my desk for signing. We have managed to recruit two experienced technicians, but not all schools are so lucky. Their responsibilities include, amongst other things, stocktaking and ordering. Clear job descriptions help them to do their jobs effectively; they also prevent teaching staff from making unreasonable demands on their time. Weekly meetings of science staff (all teachers, some technicians) help to secure the smooth running of a unit spread over a very wide area of the school.

I sign the orders and note that one is for our slowest supplier, an electronics firm. We now predict our requirements of them and order three months in

PUPIL'S NAME .... FORM .... GROUP .... DATE ....

MODULE TITLE(S) ....

DESIGNING EXPERIMENTS

FOUNDATION: I can design a simple test and predict what might happen.

INTERMEDIATE: I can design a fair test, identifying a variable and predict outcomes giving reasons.

HIGHER: I can design a fair test, identifying variables and predicting the relationship between them.

UNDERSTANDING

FOUNDATION: I have a reasonable understanding of the basic facts and ideas of the module; possibly there is some room for improvement.

INTERMEDIATE: I have shown a good understanding of a substantial portion of the work covered in this module.

HIGHER: I have shown an extremely good understanding of the work covered in this module.

EXPERIMENTING

FOUNDATION: I can make 2 or more related observations and I can use simple measuring devices in an investigation.

INTERMEDIATE: I can carry out a fair test and use suitable apparatus to observe or measure a range of events in order to produce meaningful results.

HIGHER: I can carry out measurements/observations with sufficient accuracy and can take action to control the appropriate variable.

COMMUNICATION

COMMUNICATION AND PRESENTATION
FOUNDATION: I try to present my work in a clear and organised way but I often need guidance. When doing research I include relevant information from one source.

INTERMEDIATE: I ussually present my work in an attractive, clear and logical way. When doing research I include information from different sources.

HIGHER: I always present my work in an attractive, clear and logical way. When doing research I am able to choose and acknowledge relevant information from a wide variety of sources.

EVALUATING

FOUNDATION: I can write a conclusion from my observations. I know what a fair test is and can usually explain my observations.

INTERMEDIATE: I can see simple patterns in my observations. I can evaluate my experiments.

HIGHER: I can use my patterns to predict the results of a new experiment. I understand the limitations of my experiment.

LABORATORY PERFORMANCE

FOUNDATION: I can work safely in a laboratory with guidance.

INTERMEDIATE: I can work safely in a laboratory with occasional guidance.

HIGHER: I can always work safely in a laboratory and show enthusiasm and initiative.

PUPIL'S COMMENT: (to comment on my most satisfying achievement, difficulties experienced during the module and suggestions for improvement).

TEACHER'S COMMENT: (to be discussed with pupil. May include comments on working with others and/or alone, use of equipment and materials, completing tasks (including homework), and ability to transfer skills from one situation to another).

Figure 1 High School – Science Summary Profile: Lower School

advance. There is no alternative to using them as we are equipped with several hundred pounds worth of their kit, and can afford only to repair not replace. We really should have looked more closely at this aspect of our microelectronics programme before equipping ourselves through one "cheap" supplier.

Period 2

Fortnightly meeting with my line manager, the Deputy Head (Curriculum). We discuss the progress of the faculty's review of links with FE colleges. A colleague holds a responsibility allowance for KS4 which includes these links.

I have to meet the Head next Monday to short list for a standard scale teaching post: an advertisement saying "suitable for a new entrant to the profession" produced six replies.

I report progress on the implementation of the school's new information technology policy. The move to IBM compatible computers across the school has been welcomed, but we will not give up our older BBC computers, because of the excellent control software and hardware we possess. The faculty will become the repository of the 'Beebs'. I feign disgust, while doing a mental jump for joy at the prospect of a computer in each room. My joy is tempered by the news that, next year, I will be responsible for their upkeep; repairs now take a tenth of my total budget.

Periods 3 and 4: Year 11 – Motion

Or "Off your trolley" as one of my team calls it. We structure our teaching programme to allow the best possible use of resources. We require over £2500 worth of equipment to teach our "Motion" topic and therefore have only one class set of apparatus. At present this lives on a trolley which can easily be wheeled from room to room. Increasingly, we are moving pupils and staff to rooms equipped with resources for particular topics. Close monitoring of costs over the last year suggest that "things falling off trolleys and breaking" is a major source of wear and tear. My class arrives. As they are working in small groups on their pre-determined programme, they collect their apparatus: group 1 – acceleration; group 2 – free fall; group 3 – braking distances; group 4 distance/time graphs; group 5 – "React", a game on the BBC computer designed to teach about braking distances; group 6 velocity. Light gates are set up; trolleys loaded; sand bags dropped. I spend about two minutes with most groups to check that they know what they are attempting to discover, rather longer with two of them. Practical work is reported using a standard format, designed to encourage our pupils to think in terms of Sc1. Tasks tend to be of two kinds. Explorations cover all three strands and are open ended. Pupils meet one of these every five weeks or so. The more traditional "recipe" experiments also have a place, but these are redesigned to enable pupils to develop skills in one of the three Sc1 strands. We try to organise our practical activities so that pupils can enter and exit them at a range of levels, both for Sc1 and Sc2, 3, 4. Usually we try to keep the number of different activities going on in a room to two or three, with a practical circus being the only exception. This structure has proved useful in maintaining the sanity of both teaching and technical staff. A poster on the wall reveals the "house style" for reports (Fig 2)

REPORTING EXPERIMENTS
When reporting experiments, use the following headings

Task	an outline of what you have to investigate.
Introduction	notes giving background information.
Variables	list some of the things you can alter, underline the ones you are going to investigate. which ones are you going to measure?
Hypothesis	the idea that your experiment is going to test.
Plan	an outline of your method.
Results	what happened? try to use a simple table or graph.
Conclusions	an explanation of your results. was your hypothesis right? what patterns did you see?
Evaluation	just how good was your experiment? what could you have done to improve it?
Applications	who could use your results?
Teamwork	did you work well in your group?

Figure 2 Reporting experiments

I check their homework and progress. The group and I agree, or not, that targets have been appropriately set and met. I add several notes to my records, which are a mixture of prose and grades, a more sophisticated form of the 0 – 3 scale used at KS3. Generally 2 or 3 suggests a good understanding of concepts at a particular level. Too many 3's and I am not pushing my pupils hard enough: too many 0's or 1's and I am going over their heads.

The pupils produce a range of outcomes. Ticker-tape proves useful for studying distance/time graphs, even though our computers can produce graphs more neatly and rather faster. Hands-on experience of this material still seems to help a large number of pupils to grasp the idea of velocity. Data from our invented data base reveals that it is not terribly sensible to be 16, at a party and intending to ride your new motor bike home. This message will be reinforced in our Personal and Social Education (PSE) lessons by our community liaison officer. Good citizenship is more than voting in elections and is co-ordinated across the school, like other National Curriculum themes. Science is responsible for the teaching of much of the health education theme, as well as a good deal of the environment one, in close cooperation with Humanities. Whole school planning groups ensure that these are delivered in the most appropriate context. The PSE programme is superimposed on top of the faculty structure to "backstop" the gaps and to provide a coherence that might otherwise be missing.

As the end of the double lesson approaches, I receive several pieces of work to mark from each pupil. Typically, every block of about four lesson on this topic produces two or three hours of marking. Sometimes I see rather more pieces than I mark. Some pieces will be marked against strict criteria, shared with the pupils, and carefully annotated; others will simply be marked as "seen". Our faculty marking policy suggests an approach to marking that balances the demands on the staff with the expectations of the pupils and their parents. The school's "Language across the curriculum" policy also guides our marking. Newly qualified teachers find our detailed guidelines especially useful. Quality assurance is provided by different members of the team marking identical pieces of work, and then discussing how individuals applied the policies and our criteria. We require that duplicate copies of marks are kept centrally in the science office. The IT working party is investigating ways of utilising an optical mark reader to reduce the load of marking on colleagues.

The pupils leave the room knowing that they have some preparatory work to complete for the next lesson. There is also an on-going project that they work on at home, and in the lesson if they finish early, on the design of road vehicles. The most able will investigate the theoretical aspects of momentum, but only one or two in a group reach these dizzy heights.

Lunchtime

Once a week, lunchtime is used for an informal meeting of the science staff (no agenda) when the team may review some aspect of its activity, or simply respond to a directive from the school or LEA. Once a month, formal meetings occur: these tend to be for planning. Termly, we meet for an extended period to review one or more aspects of our work, a particular topic perhaps, or the programme for a year or longer.

Lessons 5 and 6

My year 9 group this afternoon will be investigating silage! In a rural comprehensive, a rural science approach to science seems entirely consistent with the school's aims for education, expressed through our mission statement, which defines the long term goals for the school. And anyway we can get free materials from the local agricultural merchant, although the instructions on the packets tend to indicate rather larger quantities of grass than my pupils might use. ("For ten tonnes of grass..." "Yes Jo, I know your dad will let you borrow his new tractor, but the head would not allow a forty tonne silage clamp in the corner of the school field!") Increasingly, we make use of local employers, to provide both relevance and financial support.

I have asked the pupils to prepare for the lesson by doing some simple research on silage. We hold photocopies of relevant, copyright free, material that may be taken home by the pupils for planning purposes. Our notion of resource-based learning involves press cuttings in a box, just as much as the CD ROM machine in the library, presently on loan from the LEA. Resource-based learning need not necessarily prove expensive. The pupils' task is to discover the best conditions for silage production. I encourage them to work as a large cooperative group, but with each individual taking responsibility for a specific task. The level of outcome will be determined by what I see, as much as by their reports. Most of the pupils will be given the opportunity to discuss the task with me in some detail. I have to remember to take "time-outs" to note down my observations: I find it totally unsatisfactory trying to remember who I helped, and to what extent, after the pupils have gone home. Explorations get marked in detail: each piece will be annotated and given a colour grade. (I know, but numbers get boring and all our pupils know their rainbow!)

The technicians will have got out the basic materials for the lesson in advance. (Our schemes of work have detailed technician notes as well). Obviously, the apparatus provides some major hints to the pupils on how they might perform the experiment, but we feel this is fair. Each room is equipped with all the basic glassware and related apparatus, so the disturbance of the technicians during lesson times is kept to a minimum. Some explorations require that the pupils put their lesson requirements to their teacher a week in advance just as the teachers have to do for the technicians. Like some teachers, some pupils find this very difficult!

And after the lesson? As a head of faculty, I have meetings on two or three afternoons a week: my colleagues tend to have one or two. Today, I will change into my sports kit and get beaten at badminton by a fifteen year old upstart – well several actually: but it does help to see the pupils in a different context. Tonight? I am drafting an article for the ASE....

In Conclusion

Like most state schools in this country this one has spent the last three years struggling to come to terms with a science curriculum in a constant state of flux. Two major revisions in the Programmes of Study, and several minor ones as a result of dissatisfaction with a first interpretation of the National Curriculum, have left the school in a position where the science teachers can recognise good practice, but often find it difficult to implement. The progress which has been made has depended on the co-operation of employers, parents, pupils and professional colleagues: the importance of developing positive relationships with partners in the educational process cannot, in our opinion, be understated.

Throughout the changes of the last decade we have been guided primarily by one question, "How will it benefit the pupils?". The case study presented here has tried to identify some of the ways in which secondary schools are trying to ensure that pupils receive the best possible education. Not all of our lessons or our practices match the best ones discussed above but schools are increasingly recognising the importance of using a range of teaching styles in order to motivate and enthuse their customers; for customers they most certainly are, and schools that fail to appreciate this and offer anything less than a quality product, will not only fail their pupils but may end up out of business.

Secondary schools are becoming increasingly used to the language of the management consultant: "development plans", "performance indicators" and "total quality management" all feature on the agenda of management courses for teachers. These systems offer not a threat to science teachers but rather an opportunity: to develop new skills; and to take part in a revolution in the management of schools in general, and of science departments in particular. It may even be that the next edition of this publication starts with these concepts and then moves on to the classroom. Unlikely? So tell us; how do you know you are doing a good job?

Terry Parkin is a head of science in a 12–16 High School, having worked in industry and research before entering teaching. He has served on ASE Council and Assessment and Examinations Committee and is currently working on the Nuffield/Longmans' Pathway through Science programme.

Jenny Versey is Senior Teacher, Curriculum, at Alec Hunter High School. She is on ASE Assessment and Examinations committee and her Regional Committee and is a past regional officer of the SSCR.

15a Progression and Continuity in Science

Stuart Naylor and Brenda Keogh

1 What is Continuity and what is Progression?

Continuity and progression are often regarded as synonymous. This is not surprising, since efforts used to achieve one will often also be effective in achieving the other. But this is not always so. There are times when it would be more helpful to view each of them separately to ensure that they are both being successfully addressed. Just as it cannot be assumed that liaison will guarantee continuity so no assumptions can be made that continuity will necessarily lead to progression. Continuity is primarily about communication between teachers, whereas the emphasis in progression is more concerned with teacher-pupil interaction.

1.1 Continuity

Continuity is a continuous process in which all involved share views on aims and objectives, curriculum content and delivery, and methods of assessment and recording. It implies a consistency in expectation of what children can do, and in the use of suitable teaching and learning styles. Without this, problems are likely to arise as pupils move between classes or groups. These problems will be most visible moving across phases (eg infants to juniors). They will also be evident moving from year to year in primary schools and from subject to subject in secondary schools.

Teachers can continue to relate to pupils differently and employ different approaches when working with pupils to achieve common ends. Successful attempts at achieving continuity recognise diversity and build from it to enhance the learning experience of pupils.

1.2 Progression

Progression is moving forwards through a defined sequence of learning targets. There is the implication of increasing demand, challenge, complexity and achievement. However, teachers know that learning is a complex process in which consolidation and, at times, apparent regression, play an important part. Progression in learning involves teachers making day to day decisions about selecting experiences in which these concerns are recognised.

2 Why are Continuity and Progression Important

Some of the reasons include:

- to prevent duplication in learning
- to promote development in learning
- to maintain motivation, which may be the most important factor determining a pupil's level of success.
- to enable the NC to be implemented effectively by ensuring that there is a balanced coverage of its content and a common understanding of its intentions
- to help provide a clearer sense of whole school purpose, and goals which are shared.

Continuity and progression have been a long term concern for teachers and others involved in education. In 1931, the Hadow Report stated:

> *...that the process of education from five to the end of the secondary stage, should be envisaged as a coherent whole. That there should be no sharp edges between infant, junior and post primary stages, and that transition from one stage to the succeeding stage should be as smooth and as gradual as possible.*

A considerable amount of time and effort has gone into the development of schemes of work and individual programmes in science. The value of this can be seen in the improvement in the quality and quantity of science experienced by pupils in the primary phase, and in the movement towards the provision of a balanced science curriculum for all secondary pupils. However, the impact on continuity and progression has been limited.

> *The introduction of the National Curriculum has highlighted the need for improved curriculum continuity between and within schools* (HMI 1991)

Continuity and progression have tended to be viewed as an issue relating to transition between primary and secondary schools. This is where most successful teacher effort has been directed and where most research and writing has been focussed. It is easy to see why this is so. Primary to secondary transition is probably the most visible and complex situation involving issues related to continuity and progression. HMI (DES 1980) made reference to the particular importance of transfer and other writers have done likewise.

In many situations, however, the transfer between infant and junior schools is no less problematic, and from the perspective of most pupils, transition occurs annually. Once we start focussing on the curriculum experienced by the pupils (see, for example, Naylor and Mcmurdo 1990), it is not difficult to see that their concerns will include continuity within as well as between schools and progression from one learning experience to the next.

It is at the class and whole school level that we should be focussing more of our attention. This is not to ignore the importance of transition between schools. In fact, experience of the process of transition and the development of successful strategies are a useful resource to draw on to inform future progress within schools.

3 Why Hasn't it Happened Already

A lengthy list of difficulties in establishing curriculum continuity is given in Jarman (1990). It has been suggested that:

> *the reason that some teachers do not engage in such an enterprise is not because they believe it to be unimportant, but because they believe it to be impossible.* (Stillman, 1984, quoted in Jarman, 1990. p23).

However, identifying obstacles is simply the first step in finding ways to overcome them! When we begin to look more carefully at the sources of those difficulties, some of them begin to seem a bit more manageable.

(a) LEA priorities
LEA support for science has tended not to include continuity and progression and is therefore less likely to be reflected in the schools' priorities.

(b) Developing a shared view
There is still a clear need for the development of a shared view of how learning best takes place and of the teacher's and pupil's role in the learning process. The findings of groups such as CLIS and SPACE have already had a positive impact in this respect.

(c) Idealistic view
An idealistic view of continuity and progression can easily appear to demand a high level of planning, assessment and recording and contact with other

colleagues, pupils and parents. A more realistic view would set more manageable short term goals whilst not losing sight of the overall direction.

(d) Autonomy

A move towards continuity and progression can seem a threat to teachers' autonomy. Professional expertise and autonomy are highly valued. However, providing maximum support for pupils' progression in learning must surely be the most important concern.

(e) Contacts between teachers

Within schools there is a limited amount of formal and informal contact in which views on curriculum planning and delivery can be shared. In many situations, it is unlikely that teachers will observe how science is delivered in each other's classrooms. Visiting classrooms and sharing classes are not seen as the norm in many schools. Even where there is a desire to do this, there will be many other demands on the time available. Commitment from the senior management team to the potential value of contacts with colleagues will be essential, so will creative thinking about how this may be achieved.

(f) Another thing to do

Continuity and progression can easily be seen as 'another thing to do'. In fact, continuity and progression will be important outcomes of curriculum development, but only if they are made explicit as part of the process.

(g) Time management

Development implies the need for contact with pupils and colleagues. Contact takes time. There is a feeling already that there is not enough time available to achieve all that needs to be done. Few schools have made the evaluation of the use of time a priority, yet most of those who do, find that there are more effective ways of managing their time. However, effective time management has to be a central issue in making continuity and progression an integral part of whole school development.

4 How much does the National Curriculum Help?

With a centrally controlled curriculum, continuity within and between schools ought, in principle, to be more manageable.

Jarman (1990) felt that the following aspects of the National Curriculum would prove useful in promoting continuity and progression.

- It is a continuous course from Key Stage 1 to Key Stage 4
- It has common Programmes of Study and Attainment Targets
- It uses common assessment procedures relating to common Statements of Attainment
- It uses the same strands throughout the four Key Stages
- The same documents are available for all teachers.

Nevertheless, she goes on to point out that

> *'whilst it may encourage continuity, it will not ensure continuity'.*
> (p26)

The NC has highlighted the need for discussions on continuity and progression and provided the focus for them. In Bolton, for example, groups of primary and secondary teachers met to come to some common agreement on what the SoAs mean. (Bolton 1991). The need to map out what lies between some of the SoAs has also resulted in some useful developments, see, for example, Peacock (1991) or Fagg et al. (1990).

However, the NC does raise a number of challenging questions which still need to be explored:

- How does the fact that learning in science is not usually linear relate to the ten levels of attainment?
- How can the ATs take context into account? The context for level 3 at age seven is likely to be very different from that of age fourteen.
- How can the question of overload in the NC be addressed?

5 The Importance of Matching

It is easy to look at a scheme of work and feel that it will naturally lead to continuity and progression. For the individual pupil it may feel rather different. It doesn't matter what the curriculum is like in theory, it is how the curriculum is experienced by pupils that matters.

> *'Providing appropriate learning experiences... requires careful planning and sensitive teaching by teachers with a broad understanding of science and the ability to match the work to their*

*pupils' capabilities. Activities must challenge all pupils and, at the
same time, provide them all with success at some meaningful level.'
(NCC 1989 p A9)*

Matching involves attempting to get the degree of challenge and success
right – for many of the pupils for much of the time, at least. It means seeking
to provide opportunities to consolidate and extend existing ideas and skills.
Better matching will lead to better progression.

Curriculum plans describe the intended experience for the pupils.
However, what children learn depends on the interaction between the plans,
the teacher and the pupils. Matching, and therefore progression in learning,
must take this interaction into account.

Research groups such as the primary SPACE Project or the CLIS Project
have highlighted the importance of this interaction. They are currently producing materials to support teachers in anticipating pupils' ideas and
responding to them. The 'constructivist' approach which they advocate is
increasingly accepted as an effective way of teaching science.

This approach stresses the importance of recognising and building on the
learner's existing ideas. Teaching science has to be more than simply providing opportunity for practical activities and explanations to go with them. It
also has to include helping pupils to reflect on their existing ideas, to devise
investigations to test them out, to review their ideas in the light of evidence
and to realise the limitations of their ideas.

These aspects of teaching are difficult to prescribe in advance. Any
scheme of work must allow sufficient flexibility to enable the teacher to
respond to the pupils' own ideas. Continuity and progression do not simply
rely on curriculum planning.

Much science teaching already reflects this constructivist view of learning. Its
influence can also be seen in the *Non-Statutory Guidance for Science*. However,
this approach does not mean planning separate activities for each pupil, nor is it
necessary to assess everything before starting to do any teaching.

How then can effective matching be achieved? A very useful outline is
given in Harlen (1985, chapter 6), where matching is described as a dynamic
process in which the teacher attempts to adjust the activity according to the
pupils' responses.

In this view of matching, the teacher's initial plans will identify suitable
starting points and will normally provide opportunity for the pupils to share
their ideas in one way or another. It will become clear from their responses

whether adjustments are needed to make the activity more familiar, more accessible, more wide-ranging or more challenging. The teacher will attempt to reshape the activity or the learning environment where necessary, in order to ensure that pupils are given meaningful challenges which lead to development of their ideas. This is shown diagrammatically in figure 1.

The process of teaching in itself allows the teacher to gather some information to use in making judgements about pupils' capabilities, about their responses and about the learning environment provided. These judgements then provide a firmer basis for setting the right level of challenge.

Developing the skills involved in assessment and evaluation will help teachers in matching more effectively. However, pupils also have an important role in ensuring matching. Their roles will include sharing their existing ideas; learning to set their own challenges (eg through learning how to ask questions which are open to investigation); and seeing the purpose in keeping the teacher fully informed about their involvement and progress. By promoting these roles for pupils it is possible to help them to take on a greater share of the responsibility for ensuring effective matching.

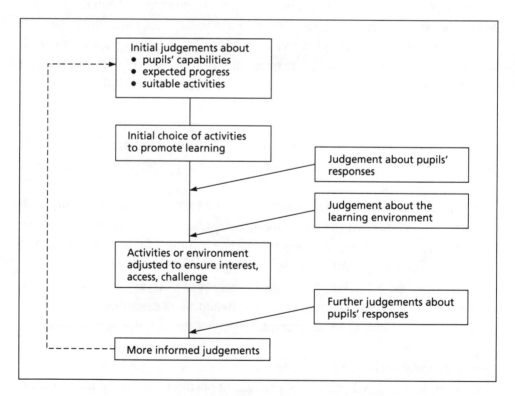

Figure 1 A responsive approach to matching (Adapted from Harlen 1985, p.144)

What about the teacher's role? The responsive approach to matching described above suggests that it will vary according to the pupils' responses, and that it will include:

- Capturing the pupils' interest and making sure that they have the opportunity to express their ideas
- Providing a range of activities and being able to adapt these in a variety of ways
- Monitoring the pupils' responses using a range of informal methods
- Providing a flexible and supportive learning environment

To a large extent these kinds of things go on intuitively. Better matching, and therefore better progression, will come from a more systematic and conscious attempt to take on these roles.

6 Progression in Scientific Procedures

Descriptions of the expected stages of the development in scientific procedures (such as those in the SoAs) are necessary for making judgements about children's existing capabilities. However, to do this is far too demanding to do in a systematic and detailed way for individual pupils particularly in large classes. A more manageable way forward would be to base decisions on a general intention to raise the pupils' awareness of how to use a scientific approach.

Very young children will often use some aspects of a scientific approach (in discovering that the soap always sinks and the rubber duck always floats, for example), but they will have little understanding of what it means to work scientifically. The long-term intention must be to get pupils to begin to understand what it means to use a scientific approach, until, ultimately it becomes an automatic way of working.

Clearly, getting the pupils "doing science" must be an essential part of raising their awareness, though "doing science" will not be sufficient on its own.

The main stages in the constructivist approach can also provide guidance for teachers concerned with the development of scientific skills. Just as clarification, reflection, challenge, application and review will lead to deeper understanding of ideas, so will they also lead to greater awareness of scientific approaches.

Teachers will need to

- Be explicit about what approach has been used during an activity
- Ask challenging questions about the approach used

- Provide opportunity to modify the approach and to apply new approaches
- Help pupils to evaluate and reflect on alternative approaches

In this way, the approach used can also be made the focus for the pupils' attention, through, for example, overt discussion of science skills.

Working in this way suggests a number of additional aspects of the teacher's role. These will include:

- helping pupils to see the need for a scientific approach
- demonstrating or explaining specific techniques when necessary, such as using a microscope or folding a filter paper
- occasionally emphasising different aspects of the procedure
- discussing effective approaches (such as examples of other pupils' work)
- referring back to the pupils' original aims and plans when reviewing
- making judgements about how effectively pupils are using a scientific approach

Many classroom activities provide valuable opportunities to practise using a scientific approach. However, the purpose of the activities must be made clear if the pupils only reflect on the results of what they did rather than on how they did it then they may end up not extending their skills at all. Doing science does not necessarily lead to learning how to do science.

7 Progression in Understanding Scientific Ideas

Planning for progression requires a vision of where we are going. A scheme of work represents an attempt to put that vision into practice, so that progression in understanding will result.

In the same way that the National Curriculum attempts to describe the expected stages in the development of scientific skills, it also attempts to do this for scientific ideas. The NC does not really provide the kind of vision that teachers need. The steps between the SoAs are generally too large to be tackled in one go, so that a single step will normally have several activities associated with it.

Some useful recent work describes, in much finer detail, the possible stages in the development of scientific ideas. Peacock (1991) outlines a detailed sequence of ideas in the area of floating and sinking. CLISP are currently producing a series of guides to describe the progression in scientific ideas at Key Stage 3.

Although these sequences of ideas are invaluable, they also need to be balanced by other perspectives. They need to be assimilated for use on a day-to-day basis in responding to the pupils' own ideas. This raises a serious question: how are teachers to respond when they are unable to assimilate these detailed sequences or when such sequences are not available? Pupils will still have their own ideas and it will still be necessary to make some kind of response to them.

A general framework will help to provide an overall view of how scientific understanding might develop and support decisions about how to respond to the pupils' own ideas. A suggested framework is shown in figure 2.

The section on matching indicated the importance of teachers adapting activities to make them more accessible or more challenging, according to how the pupils initially respond. This general framework provides some guidance on how to adapt activities in this way. It gives a broad indication of the type of activity which is likely to be appropriate for pupils at various stages in their learning, in any conceptual area.

Whatever the age of the learner, exploring will precede investigating which will precede researching . Younger pupils will be encouraged to concentrate on experiencing and exploring phenomena; this is an essential stage in their early development of scientific ideas. Further challenge can be provided by supporting the pupils in attempting to find explanations for their experience and beginning to investigate their ideas more systematically.

Older pupils will normally be encouraged to place a stronger emphasis on systematic investigation and research. However it would be a mistake to ignore the need for exploration. Even teachers will want to 'play' with new materials, equipment or ideas! One important feature of this general framework is the way that it highlights the distinction between observable and non-observable events. Many early scientific concepts will emerge naturally through experience (the idea of solids, liquids and gases for example). Through investigation, these ideas can be refined and explanations sought which are consistent with the evidence available.

However there is a limit to what the pupils can discover on their own. Once we enter the realm of more abstract ideas which are not directly observable then investigation becomes insufficient. Abstract ideas, such as electrons or gravitational force, are created not discovered; we cannot simply expect pupils to discover these complex concepts for themselves. The teacher has an important role in providing access to these new ideas in the most appropriate way.

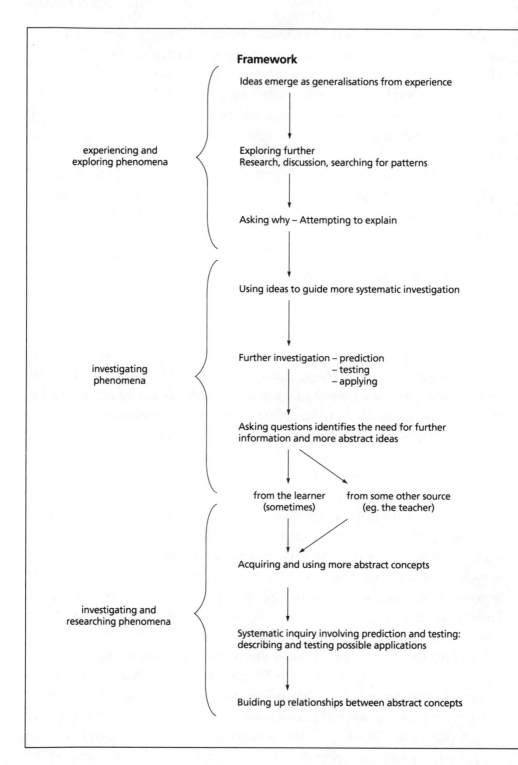

Framework

Ideas emerge as generalisations from experience

experiencing and
exploring phenomena

Exploring further
Research, discussion, searching for patterns

Asking why – Attempting to explain

Using ideas to guide more systematic investigation

investigating
phenomena

Further investigation – prediction
– testing
– applying

Asking questions identifies the need for further
information and more abstract ideas

from the learner
(sometimes)

from some other source
(eg. the teacher)

Acquiring and using more abstract concepts

investigating and
researching phenomena

Systematic inquiry involving prediction and testing:
describing and testing possible applications

Buiding up relationships between abstract concepts

Figure 2 Progression in conceptual understanding: a general framework...

Some illustrations

Something happens – I see what happens or someone helps me to see I look more closely – Have I seen what is really happening?	Ice melts It feels cold, it feels wet... the outside melts first
Does it always happen the same? Can I change what happens or can I see something make it change?	Sometimes it melts faster I can warm it up I can wrap it up
Why does it happen? Can I explain what I think?	When it was near the radiator it melted faster When I wrapped it in the newspaper it did not melt as fast
What do I think makes a difference? Can I change what happens in a systematic way so that I really find out what matters?	It seems to be something to do with being warmer or stopping it getting warm I can measure the temperature, the time, etc. I can try to identify variables
Can I try it in other ways and get the same pattern of results?	I will use bigger blocks of ice/higher temperature/ lower temperature/different insulation, etc
Can I explain what is happening in observable terms?	The higher the temperature – the faster it melts The greater the volume – the slower it melts More insulation – slower melting
Is there more I can find out or you can tell me to help me to understand more about why it happens?	You tell me about particles. I read a book with more information about solids, liquids and gases (I am unlikely to discover atoms on my own)
Can I use what I know to explain what is happening in non-observable terms?	I can now explain melting using what I know about particles and explain why the temperature makes a difference
Can I use the idea to explain other changes?	I realise that condensation, freezing and evaporation all use the same theory
Can I use other ideas to explain a greater range of events?	I can explain ice floating (volume and density) and think about the possible arrangement of particles

and some illustrations

Progression in the development of these ideas will not be achieved simply by providing "the theory". It will rely instead on modelling, predicting, testing, interpreting, reviewing and applying theoretical ideas in a meaningful way. The skills of scientific investigation therefore play a vital role in the development of scientific ideas.

8 Practical Ways Forward

When making decisions, it is important to be realistic. No one can do everything and certainly not by tomorrow! Like matching, progression is an ideal to aim for rather than something which is completely attainable. We have identified below some of the things that can be done to improve the experience of many pupils.

8.1 What can you do with your own class
Recognise pupils' capabilities
Recognising what pupils are able to do for themselves or for each other is important in enabling you to use your time most productively. Identify something which you feel the pupils may be able to do, think through any implications for your classroom organisation, then let the pupils have a go. Don't expect things to be perfect first time – but it will get better.

Identify the pupils' previous experience
When starting work on a new area, spend time with the class discussing their previous experience in that area. This should help avoid repetition and help to give you some indication of what pupils' ideas are.

Encourage feedback from pupils
At the end of sessions, ensure that there is time for feedback from the pupils. Take this as another opportunity to listen to and challenge their ideas; to get them to justify their ideas; to question each others' ideas, to raise new questions to be explored; and to redirect or refocus if necessary.

Reflect on your learning
Reflect on something you have learnt in science recently. What strategies did you use? How do these strategies compare with the ones you are using with your pupils? Have you asked your pupils what helps them to learn best?

Encourage pupils to reflect on their own learning
At regular intervals encourage the pupils to look back over their work and reflect on what they have achieved. This will help them to understand the purpose of the activity and the criteria for success, so enabling them to take more responsibility for their own learning.

Provide opportunities for demonstrating learning
Give the pupils as many opportunities as possible to show in some tangible way what they understand and can do. A written report may be the most appropriate way, or asking pupils to think about issues, to answer questions, to provide annotated drawings or posters, to produce a concept map or summarise how their views have changed.

Provide guidance for recording
Making thinking visible is not always easy, particularly in relation to scientific skills. Structuring the pupils' recording may help to clarify what their thinking is. Recording sheets such as those for the 1992 KS 1 SATs for Sc1 may be useful. The intention would be to give the learners the best opportunity to show their current understanding of scientific skills.

Record experience
Achieving continuity in learning relies on knowing something about the kinds of experiences pupils have already had and the point at which they stopped learning. You may find it helpful to have separate sheets of paper for each of the strands in science on which to briefly jot down the main experiences provided for the class. It will be helpful to identify the point at which learning ceased and to make a note of individual pupils who struggled or advanced further than the majority of the class.

Note significant points in the pupils' learning
Speech bubbles stuck into their books or included in folders can be used by you or the pupils to record points of significance to return to at a later stage. This will help the pupils to share the responsibility for follow-up, as well as keeping the comments in the most useful place.

8.2 What can you do with the other teachers?

Contribute to discussions

Ensure that your ideas about science are heard. Continuity and progression depend on each teacher's recognition of the approaches of the other teachers in the school and on beginning to work towards common approaches.

Ask questions of other teachers
Ask questions of the previous teacher, eg:

- Which areas of science did you cover?
- Which did the pupils find most difficult/easy/most interesting?
- Which approaches best motivated the pupils?
- Which aspects of skill development need help?
- How independent are the pupils. How much responsibility are they able to take?
- What kinds of records have been kept? Which of these were most useful?

Provide information for other teachers

Have the same kind of conversation with the teacher who will be receiving pupils from you. This time, focus on the type of information which you will be able to pass on. What information will the next teacher find helpful? Are you using a system which gives you what you want and which is able to effectively inform the next teacher?

There are two purposes for these conversations. One is to enable you and the next teacher to have a clear picture of the pupils and their experiences. The other purpose is to enable you to have discussions about, and therefore raise awareness of, issues related to continuity and progression.

Work alongside your colleagues

If you have some non-contact time available, try to negotiate to visit other teacher's classrooms and have colleagues visit you. During these visits you could look at:

- how science is taught in other classes, particularly those before and after yours
- particular approaches or ideas which you would like to try, perhaps with a colleague

- classroom organisation and teaching styles
- the next group of pupils you will be working with.

The important principle is to work with colleagues towards common approaches to science learning and the roles and expectations of the children.

Progression within a class and continuity within a school are fundamental concerns of education. Problems at transition may happen to be the most visible, and while these are important they should not be seen in isolation. The purpose of liaison is to lead to better continuity; the purpose of achieving continuity is to lead to better progression in learning.

Stuart Naylor and Brenda Keogh work in the School of Education at Crewe and Alsager College of Higher Education and Manchester Polytechnic respectively. They both have experience as primary and secondary teachers and as laboratory technicians; and share an interest in in-service teacher education. Currently, they are members of the ASE Working Party on Progression and Continuity.

References

Bolton LEA (1991) *My level 3 – your level 3?* Bolton LEA

CLIS Project (1987) *A constructivist view of learning and teaching in science.* University of Leeds

DES (1980) *A View of the Curriculum.* HMSO

DES (1989) *Science Key Stage 1 and 3.* HMSO

Fagg, S., Skelton, S., Aherne, P. and Thorber, A. (1990) *Science for All,* Fulton

Hadow Report. Great Britain Board of Education Consultative Committee (1931) The Primary School. HMSO

Harlen, W. (1985) *Teaching and Learning Primary Science.* Chapman.

Harlen, W. (1992) *The Teaching of Science.* Fulton.

Jarman, R. (1990) *Primary science – secondary science continuity: a new ERA?* School Science Review Vol 71 No 257 pp 19-29

Leeds City Council/CLIS (1992) *Leeds National Curriculum Science Support Project* Leeds City Council/University of Leeds

Marshall, P. (1988) *Transition and Continuity in the Educational Process.* Kogan Page

NCC (1989) *Science Non-Statutory Guidance*. National Curriculum Council

Naylor, S. & McMurdo, A. (1990) *Supporting Science in Schools*. Breakthrough Educational Publications

Peacock, G. (1991) *Floating and Sinking*. Sheffield City Polytechnic/NES Arnold

Secondary Science Curriculum Review (1984) *Primary Science, Secondary Science: some issues at the interface*. SSCR

Secondary Science Curriculum Review (1987) *Better Science: Building primary – secondary links*. Heinemann/ASE

SPACE Project (1992) *SPACE Teachers' Handbook*. Collins

Stillman, A. and Maychell, K. (1984) *School to School: LEA and teacher involvement in educational continuity*. NFER Nelson

Tickle, L. (1985) *From Class Teacher to Specialist Teachers: curricular continuity and school organisation* in Derricott, R. (Ed) *Curriculum Continuity: primary to secondary*. NFER Nelson

15b Progression and Continuity in Science
The Norfolk Experience

Brian Betts

A document well worth revisiting in the light of current developments is *"Science 5–16: a statement of policy"* (DES 1985). It marks an important turning point, for this made a clear statement that primary science must be a part of the curriculum in all schools.

It also put the issues of continuity and progression firmly on the map, with a clear statement about the need for liaison and continuity between the primary and secondary phases. It pointed to the need for mutual recognition of the special responsibilities of each phase and the need to give value to the work undertaken by partner schools.

For many years primary school science was a peripheral activity, often lurking in the guise of 'nature-study', and regarded very much as an add-on activity. This is not to decry the value of the study of natural history, but rather to highlight the methodology employed in its teaching, which paid little heed to the processes of inquiry and experimentation and generally presented a collection of rather indigestible facts. However, a number of enthusiastic primary teachers with an awareness and understanding of the importance and relevance of science in the curriculum, developed process-based work, often stemming from the 'nature-study' activities, which made important contributions to the education of the children in their schools. These cases were, however, in the minority and had little impact in terms of continuity between primary and secondary phases, there being a generally held view that science education did not begin in earnest until children reached the secondary school.

Science 5–16 provided a guide for much of the work done in Norfolk during the period from 1985 to 1990, when a primary science project team worked across the county supporting schools in discovering and understanding the exciting nature of investigative science. Because of the mix of infant, first, junior, primary, middle and high schools, liaison and continuity were vitally important. Local support groups were set up, and these were very effective in encouraging liaison activities. Often, groups would elect to produce their own local schemes of work and set up centres, in one of the schools, to pool and share teaching resources and equipment. Funding towards this was, when possible, provided by the project team. The idea of the 'science topic box', containing all the necessary items for a particular theme, proved to be popular. Team members were very active in schools and with local groups.

Secondary schools were keen to participate but were often concerned about the coverage of science content. To help in this, some set up INSET sessions for primary colleagues within their areas. Secondary schools recognised the value of the work on the processes of science in the primary schools and felt that the heightened levels of science understanding and skills achieved would be a major bonus to their own work, particularly in the first two secondary years. There was some way to go to reach an overall consensus regarding the content element, and it is not unreasonable to say that there was a perception amongst some colleagues that the science content should not be seriously tackled until children reached secondary schools. This should not, however, detract from the overall effectiveness of this period of development and the excellent work done by so many to build new relationships and share professional expertise, in the cause of science education and for the welfare of children.

With the advent of the National Curriculum, these early experiences and efforts proved to have been of enormous benefit. They enabled schools to look at the requirements of the NC and, with liaison structures and networks already in place, to work to the common format which it provides. One of the great benefits of the Science National Curriculum is that it offers a basis for a continuum of science education from 5 to 16 years, both in terms of investigative work and content. It provided an additional impetus to the work of local area groups, removed any sense of 'divide' between phases and gave a greater sense of involvement to all concerned. As schools have continued to work together, an important feature which has developed has been the element of equality of partnership, with colleagues from all types of schools and phases receiving equal, mutual regard.

A particular difficulty which needed to be addressed was the mismatch between the Key Stage age bands and those of the Norfolk schools: first schools 5–8yrs., middle schools 8–12 yrs., primary schools 5–11 yrs., and high schools either 11 or 12-16yrs. In some cases, high schools receive pupils from more than twenty different middle schools. In other cases, the situation is more manageable, with partner schools all being within the same catchment area. A number of high schools receive pupils at age 11 and at age 12.

There was a clear need to provide some advice to help with this situation so, through a working group, in close liaison with a large number of teachers, Norfolk Science produced two documents to meet these needs. Both have since been updated to the new (1992) Science National Curriculum.

In outline the documents work in the following way:-

- For Key Stage 2, sections of Sc2, Sc3 and Sc4 Programmes of Study have been allocated to year 3 in first schools and to years 4, 5 and 6 in middle schools.
- Similarly, for Key Stage 3, sections of Sc2, Sc3 and Sc4 Programmes of Study have been allocated to year 7 in middle schools and to years 8 and 9 in high schools.
- The Programme of Study for Sc1 has been given equal emphasis for all years in each Key Stage.

A further document entitled '*A Guide to Primary Science Development*' has been written which covers most aspects and concerns of incorporating science into a school's primary curriculum. This was also produced as a result of close liaison between teachers from the primary and secondary sectors. These documents are widely used in Norfolk schools and have proved to be very effective in support of liaison and in establishing progression and continuity in science education.

Brian Betts is General Advisor for Science in Norfolk LEA.

References

Science National Curriculum: Key Stage Two – A Document for Guidance and Liaison. 1992. Norfolk Education Authority

Science Key Stage Three: Norfolk Science Non-Statutory Guidance. 1992. Norfolk Education Authority

A Guide to Primary Science Policy. Norfolk Education Authority 1992

Science 5-16 a statement of policy 1985 DES/WO.

16 Science at Ages Sixteen to Nineteen

David Billings

1 Introduction

The current education provision at 16-19 is extremely wide-ranging, highly fragmented and inconsistent across the country. There are small and large sixth forms, sixth form colleges, FE colleges, tertiary colleges and independent schools, all working under a variety of regulations. There is also a wide array of courses in both the academic and vocational fields. In many cases institutions teach these with little or no knowledge of or reference to other providers. This can all be very confusing to pupils, parents and employees.

Some important developments have sought to alleviate this problem, in particular the TVE initiative and its extension programme which have encouraged the development of partnerships and consortia among the 16-19 institutions. But many feel that the curriculum is still divided between the academic and the vocational, and that there is a need for it to be more unified under a national qualification system which recognises each route as equal. (However these issues are still contentious.)

What then are some of the requirements? Post-16 provision should accommodate the needs of all the cohort, not just the most able. The curriculum should offer a broad and balanced system of qualifications which enables students to keep their options open, to avoid early specialisation. The curriculum also needs to be more relevant to the needs of employment. Consequently, programmes should include core skills and involve employers in the provision of a greater range of work-related skills, developed both on work experience and in work-based assignments. This implies a greater exchange of experiences between the worlds of work and education than has been the case in the past. At the same time the system needs to build onto the GCSE provision at one end and be acceptable for entry to higher education at the other.

What then is happening to improve the situation?

In the academic field, the attempt to produce an increase in breadth and balance has been by the modularisation of A levels and the development of

AS levels, (p 300), although this provision has perhaps not been as success-ful as anticipated.

The work of the National Council for Vocational Qualifications (NCVQ) has extended this development into the vocational field. But as we shall see, this work is likely to have an impact on the whole 16-18 scene and lead to a rationalisation of the system.

2 NVQ's and GNVQ's

The National Council for Vocational Qualifications was set up by the gov-ernment in 1986 following proposals put forward in a White Paper to review the system of vocational qualifications, in order to produce a rationalisation. It aims to bring all vocational qualifications into one national system. It has proposed two forms of qualification, National Vocational Qualifications (NVQ) and General National Vocational Qualifications (GNVQ). NVQ's are very much occupationally specific and there is not space to consider them here. How the levels of NVQ qualification relate to other qualifications including GNVQ is shown later (p 295) More important from the point of view of science education are GNVQs.

2.1 GNVQs
The aim of GNVQ's, as specified by the governments' White Paper *(Education and Training in the 21st Century)* is to provide a broad based vocational education as a genuine alternative to A level qualifications, but of a comparable standard. GNVQ's are designed to provide both the acquisition of basic skills and an understanding of the underpinning knowledge in a voca-tional area, together with a range of appropriate core skills. Also, more impor-tantly, GNVQ's must provide this broad-based alternative within, and aligned to, the frameworks of NVQ's, GCSE's, A levels and degree programmes. Since GNVQ's are set to form the basis of future curricular activities it seems sensible to spend some time in discussion of their mode of operation.

At the time of writing (July 1992) NCVQ have developed, and are set to pilot from September 1992, GNVQ's in five broad areas. These are manufac-turing, art & design, leisure and tourism, business and health and social care. Although science is not one of these areas, it has now been agreed by the National Council that science will be in the second phase group to be devel-oped from September 1992, with piloting to take place in schools and col-leges from September 1993.

GNVQ's offer a unit based scheme at different levels, corresponding to those at NVQ. (see page 295) A GNVQ at level 2 is equivalent to a GCSE, and one at level 3 to GCE A level.

Level 3

In order to achieve a GNVQ at level 3, a student must complete 12 units. Of these 12 units, 8 are mandatory and 4 chosen from a list of options. A GNVQ at level 3 with 12 units is then equivalent to 2 GCE A levels.

A student may study an additional A level alongside the GNVQ units, or more able students can be encouraged to study 6 more units taken from an additional list. The 18 unit GNVQ is then equivalent to 3 GCE A levels, and provides a satisfactory progression on to higher education. The additional units studied may include not only vocational units but also such units as foreign languages and mathematics, particularly if these are required to gain entry to a degree programme.

In addition to the above units, students will be required to achieve 3 core skill units at the specified level, these being in communication, application of number and information technology.

Level 2

GNVQ's at level 2 will consist of 6 mandatory units, which are taken from and relate to the mandatory level 3 units. They are, however, less demanding than those at level 3 and will deal with basic skills and principles. It is essential that, since level 2 GNVQ's are equivalent to GCSE, they relate properly to the National Curriculum, and this is explicitly stated in the characteristics of GNVQ's as put forward by the NCVQ.

> *.... it is also proposed to ensure that a clear relationship should exist between level 2 awards and attainment in the National Curriculum. It is anticipated that level 2 will be broadly equivalent to National Curriculum level 7 or slightly above. The relationships between statements of attainment in the National Curriculum and those in general NVQ's will be explored in detail to maximise alignment between the two systems.*
> *(NCVQ 1991 section 4)*

It is anticipated that GNVQ's at level 2 will take one year of full-time study whereas GNVQ's at level 3 will normally take two years full time. However there is an in-built flexibility in the scheme, for them to be offered over longer or shorter periods and through part-time and open learning study.

The units in each GNVQ are formulated by a suitable development group consisting of relevant institutions together with awarding bodies. The above groups will identify the unit titles within the GNVQ, and then each unit will be specified in terms of statements of achievement (outcomes). For each statement, evidence indicators will be provided, giving performance criteria by which assessments will be made.

The GNVQs will fit into a structure with all the other programmes. This structure is indicated in figure 1. It can be seen that progression is through GCE A levels and GNVQ levels on the one hand, and through occupationally specific NVQ levels on the other. There is also a clear attempt to organise the qualification system into a rationalised coherent whole and to bridge the so-called academic – vocational divide. This can only be achieved if all the qualifications are linked in terms of accessibility, progression and level. This will enable not only progression within 'routes' but also movement between the academic and vocational programmes, as for example where a student may

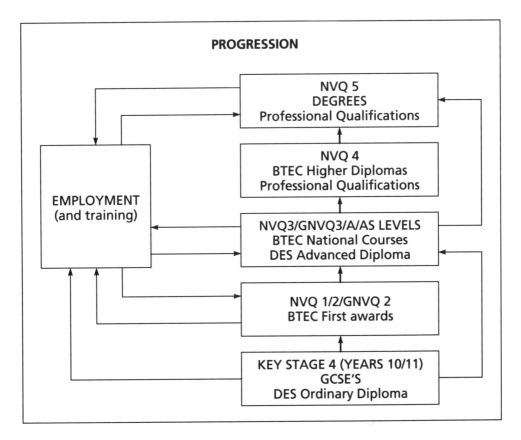

Figure 1   The future structure of post-16 programmes of study.

progress from a GNVQ level 2 to a GNVQ level 3; or to employment and training and on to a NVQ level 3. This will be possible because everyone will know, including industry, what a level 2 or 3 qualification represents, whether it is in GNVQ, NVQ, or A level. At last a coherent and rational system!

2.1.1 Teaching and learning strategies in GNVQ

The teaching and learning strategies in the new GNVQ's can be inferred from the way in which NCVQ see the assessment.

> *The primary form of evidence for assessment will be derived from projects and assignments carried out by a student. It is expected that students will take a leading role in assessment by collecting and presenting evidence to ensure that they cover, and show that they have covered, all the requirements of the elements and performance criteria in respect of each unit.*
> *(NCVQ)*

Irrespective of this statement of intent, one would be automatically led to the same conclusion by the way in which the elements within the units' general Statements of Attainment are presented. Since, as yet, no science GNVQ exists, it is obviously not possible to give a specific example of the Statements of Attainment required. The example below is provided as a nearest equivalent. It is given by NCVQ as part of the mandatory unit "Develop Design Specifications" in the Manufacturing GNVQ programme at level 3. (NCVQ 1991 Appendix B) It can be seen that the student must produce a written project providing a design specification. This could obviously involve the student in gaining information, through correspondence, interviews etc. and in the evaluation of its worth.

Example of a GNVQ Unit at Level 3 (Manufacturing).

Unit 1 Develop Design Specifications.
1.1 Specify design criteria.

(a) *sponsor requirements are clearly and accurately elicited and summarised.*
(b) *summary requirements clearly distinguish between technical and aesthetic criteria and identify priorities.*
(c) *technical requirements are accurately described and related to relevant codes of practice, conventions, output requirements and operational features.*

(d) *relevant design, manufacturing and operational standards are accessed and incorporated into the design criteria.*

(e) *the complete design criteria are presented in a recognised and legible format.*

Range

Eliciting requirements:

direct (e.g. face to face interview)
indirect (e.g. written commission, market analysis).

Technical requirements:
(i) *physical characteristics (e.g.shape, size, materials, finish/colour/appearance and dimensions)*
(ii) *output characteristics (e.g. costs, quantities required, schedules, recognised aesthetic standards and conventions)*
(iii) *constraints and limitations (e.g. manufacturing system limitations and capabilities, legal regulations and restrictions).*

Source of standards:
Statute (e.g. British Standards, European Standards, statute law)
non-statutory (e.g. codes of practice, conventions).

Evidence Requirements:
A written design specification incorporating all the characteristics in c and d across the range of technical requirements. Requirements identified from written specifications and an observed interview with a 'sponsor'. Supplementary testing of sources and types of standards.

It is worth noting some other aspects of the assessment criteria laid down in the GNVQ documentation. Students will build up a portfolio of evidence in support of their programmes which will be used to form a part of the national record of achievement as well as providing evidence for the assessment and grading of the GNVQ units. It is also possible that students can build up evidence from experiences outside the GNVQ programme. This will be particularly so in the assessment of core skills, where work outside of school or college may be used. Alternatively, evidence obtained from prior learning or courses, even perhaps at GCSE level, may be applicable.

This build-up of evidence for assessment purposes will be supplemented by unit tests, which will tend to be short ones, set at appropriate points in the unit, rather than the more usual long, written examinations at the completion. The unit tests will be externally set by the awarding body.

The style of delivery of the GNVQ could be vastly different from that normally experienced. The achievement of the outcomes is stated in terms of projects and assignments. Industry is to be used as a source of information, to be gained from visits and interviews. Where possible the projects and assignments must be linked to the world of work, through some form of work experience, or, if this is not possible, through the use of simulations and/or case studies which can give students a glimpse of the industrial world.

3 BTEC

But, stop a moment or two. Is all this 'new way' of learning and assessment totally revolutionary? Have we not come across these methods somewhere before? Let us have a look at other programmes within the proposed educational framework and see how it is thought best that these are delivered. Certainly, those centres with experience of delivering BTEC courses have the necessary background in involving students in active learning. BTEC actually encourages this type of learning by formulating a set of criteria which centres must fulfil in order to achieve the necessary quality rating to satisfy BTEC that a course is operating satisfactorily. These criteria can be summarised as follows:

a) In terms of course delivery
A method of course delivery is determined that reflects the knowledge, skills, understanding and application required by the students. Student learning activities are to be organised to focus on systematic and integrated development of vocationally relevant knowledge and skills, with the timetable being organised to allow the course to be delivered in an integrated manner, not with each unit completely separate with no reference to other units within the programme. The relevance of the course is to be achieved by an active strategy to forge and maintain links with employers.

b) In terms of teaching and learning strategies
The course must provide opportunities for students to be actively involved in and given responsibility for their own learning, through the use of student-

centred learning activities. These activities need to be planned to bring about the integration of learning both within and between units, by the identification of core themes which should run across several units. An example is shown in appendix 1. This assignment on energy is one in which contributory units would be expected to include physics, chemistry, biology, mathematics and I.T. Practical work and assignments should provide work-related and/or work-based situations relevant to present or future employment. They would place students in appropriate roles and use realistic support materials. This relevance can only be satisfactorily achieved through consultation with employers and their involvement in the design of both practical work and assignments. It would also be advisable and beneficial to the student if some sort of work placement or simulated work experience could be designed as a learning activity, and this could even be used as part of the assessment strategy for the programme.

c) In terms of an assessment strategy
BTEC recognises that, although a variety of assessment methods should be employed, there should be a balance between the formative and summative, and the assessments should be realistic and relevant to work-related activities. With BTEC units at most centres, a particular unit may be assessed via the use of projects, written assignments, practical work, in-unit tests, and end-of-unit tests. However this can only be satisfactorily achieved if the student is fully aware of the aims and expected outcomes, as shown in the example given appendix 1.

4 GCE Developments

4.1 GCE A level
The future of A levels has been much discussed in recent months, particularly the issue of whether the government will put forward proposals to change their "gold standard" traditional A level format. This is seen by many as desirable to fit in better with the revamped GCSE's now starting. An ASE report commissioned in May will indicate, as have other reports, that although the number of pupils studying double award GCSEs is increasing rapidly, the numbers going on to take A level physics are dropping and to take chemistry are, on average, staying static: only in biology are they increasing.

The indications are that A levels will stay. SEAC have issued new guidelines for A and AS examinations with the agreement of the Secretary of State

> *"in order to provide a framework for the continuing development of a system of A and AS examinations that fully maintains the standards currently associated with A levels" (SEAC 1992)*

At the present time a pilot exercise is taking place to develop a core for AS and A level syllabuses. This aims to establish the knowledge, concepts and skills that pupils need to build from GCSE to A levels and beyond. A number of subject groups have already reported to SEAC. It remains to be seen what will emerge from this work.

Two recent developments in GCE A level programmes are of particular interest since they do, on the academic front, move some way towards closing the gap in teaching, learning and assessment strategies with the developments in vocational programmes. These are the Advanced Supplementary (AS) examinations and the modularisation schemes which have been produced in the last few years.

4.2 GCE AS level

The Advanced Supplementary syllabuses were introduced following a recognition that a lack of breadth was experienced by those students who studied a programme consisting of just 3 GCE A level subjects. The AS examinations are designed for a two year programme of study with a syllabus of half the content of an A level, but at the same level. This means that students can study a mix of both A and AS subjects. This introduces some breadth but arguably not enough.

4.3 GCE Modular schemes

A number of modular schemes have been or are being developed. Among these are the Wessex Project and the Cambridge Modular A level science project, both being developed in collaboration with examining boards, the Wessex scheme with the Associated Examining Board and the Cambridge scheme with the University of Cambridge Local Examinations Syndicate. All the modular schemes have the common theme of being able to provide students with a choice of modules, leading, it is hoped, to greater motivation and interest in A level science. This should go some way towards alleviating the problem of reduced student uptake of A level sciences. Modular schemes are seen as building on pupils' experiences of GCSE integrated, balanced science courses. The aims of the schemes show the intended merits of these developments.

- to provide courses which are both interesting and stimulating;
- to allow smooth transition from pre-16 science courses;
- to provide continuity with GCSE through assessment and teaching styles;
- to broaden the base of A levels by incorporating NCC skills and themes;
- to provide modules which foster a knowledge and understanding of science in the real world and the world of work;
- to emphasise a student centred approach with active learning, encouraging the development of a range of study skills both oral and written;
- to provide a good scientific background for those students who intend to progress onto further or higher education;
- to apply short term assessment targets within the modules.

Most schemes revolve around the provision of a core which can, as in the case of the Wessex scheme, be as much as 60% of the course. The other 40% consists of a number of complementary modules chosen from a module list. The in-built flexibility of the schemes means that the study of core modules in one science area does not pre-determine the choice of complementary modules, and it is possible to study modules from other subject module banks. The flexibility also allows for a combination with other programmes; for example there can be modules common to A level and BTEC programmes. This allows bridges to be built between the academic and vocational sides which have hitherto been very separate.

It can be seen from these schemes that the move is very much towards the use of a comprehensive range of teaching and learning and assessment styles, much along the lines of those envisaged in the aforementioned GNVQ's, to enable students to increase their ability to think, to work on their own, and to communicate effectively. It is likely that the SATIS 16-19 material (p 352) will be extremely useful for such courses, as they will for science GNVQ's when they come on stream.

5 Resource based learning

The issue of the development of active learning styles runs right across the spectrum of post-16 programmes. So how might this increase in active learning be achieved? One important issue is the development of resource based packages. Good ones are quite difficult to produce, not least because one of the prime aims must be for the students to be able to see clearly not only what they are going to achieve but also how the particular package fits into the whole scheme. They can vary in size from those intended for a complete module to those which will cover just a part of a module or unit, or even a single individual lesson. Each package needs to include a number of activities in it, and may well refer the student to other sources of information or resources. For example, it could include exercises, practical work and the use of I.T. A package needs to ensure that the student understands the curriculum, so it may be that the teacher needs first to discuss with the student what has to be learned and why, in order to meet the objectives of the package. It should also be designed to help each student to set his/her own learning targets and this may mean that students will be learning at a different paces from each other. The teacher in these situations will be expected to act as a guide and facilitator and, above all, to support the student through his learning process. It is also vitally important that, at the end of the learning package, a review forms part of the active process, so that the students know that they have achieved the stated outcomes and can identify their own strengths and weaknesses.

There must be a satisfactory organisation of resources, not only so that students can make effective use of them but also so that they have access to them at the times when they are required. "Resources" would here include not only packages and laboratories but also other areas of the school or college, such as the library and computing facilities.

This brief discussion shows that there is a lot involved in flexible learning. A recent Employment Department publication on the subject says:

> *The operational framework (for flexible learning) consists of three interlocking components directly linked to the learning cycle. Each component implies a change in emphasis within the teaching/learning situation, from teachers presenting information and directing student activity to teachers facilitating student action plans and their implementation in pursuit of more effective learning. This means that many teachers will need support as they change the emphasis from teaching to learning. (TVEI Unit)*

The three interlocking components mentioned above are stated as the managing of teacher/student partnerships, managing student learning pathways, and managing student use of resources. They are all equally important for a successful application of a flexible learning strategy. Perhaps one final point is worthy of mention in terms of the third of these components. If we are to pursue a flexible learning approach, it is imperative that flexibility is built into the timetabling and rooming to allow students uninterrupted access to resources and/or laboratories. This, of course, can be a major problem, and different schools and colleges have tackled it in different ways. Perhaps one of the more common is the development of open access workshops. These are areas which contain the flexible learning packages, I.T. facilities and other relevant resource information. They can operate in a number of ways. For example, all classes can be timetabled into them as part of their curriculum, on an open access basis so that students can book into them individually. During their class time within the workshop, students will be working on a particular package forming part of a unit or integrating a number of units, as described in the previous pages. An important point is that the workshop needs to be staffed at all times so that students can get help as and when they need it. It is a concept certainly worthy of future investigations.

6 Conclusions

In this chapter I have tried to give the reader a feel for the current practice in science 16-19 and what developments are imminent. The major development is in GNVQ's, with the attendant emphasis on a more flexible approach, giving students a greater breadth in their programmes and more responsibility for their own learning. This trend will, I feel sure, spread across the whole of the 16-19 curriculum, giving a rationalisation of qualifications, an enhanced progression from pre-16 (GCSE) to higher education, and possibly a closing of the academic / vocational divide, which can only be to the good. Those of us working in this sector have the challenge of putting some of these ideas into practice ready for the onset of such changes as GNVQ's in 1993. We live in interesting times!

David Billings is Head of Mathematics and Science at Peterborough Regional College. He is a member of the ASE Post-16 Committee with particular responsibility for NVQ matters. He has lectured in physics on a variety of courses up to degree level.

References

ASE (1992) *Research into A level science uptake*
University of Cambridge Department of Education. *Cambridge Modular Sciences Project*

DES (1988) *Advanced Supplementary examinations: The first two years.*

NCVQ (1991) *General National Vocational Qualifications: Proposals for the new qualifications.*

SEAC (1992) *Principles for GCE Advanced and Advance Supplementary Examinations*, SEAC.

Tomlinson P & Kilner S *The flexible approach to learning: A guide.*

TVEI Unit, Employment Department. *Flexible Learning: A framework for education & training.*

University of Sussex Institute of Continuing & Professional Education. *Flexible learning in schools.*

The Wessex Consortium. *The Wessex project: Modular developments post-16.*

Appendix 1

BTEC 1st Diploma in Science

Assignment number 4 – Energy.

Aim
To allow the student to examine the current sources of energy and, bearing in mind the finite quantities of certain resources, offer solutions to provide for our long term energy needs.

Common Skills Assessed
Investigation, Self Organisation, Communication, Problem Solving, Analysis and Numeracy.
NB. I.T. must be applied where appropriate

Course Integration
Expected contributory units include:
   Core science
   Physics
   Chemistry
   Biology
   Mathematics
   Information Technology

The Assignment Itself

1. *Draw and describe the operation of an electrical generator. Draw and describe three ways of producing the rotary motion required by a generator.*

2. *Write to different organisations requesting information about energy sources for power generation. Include some of the information with your submitted work.*

3. *For the fossil fuels produce a bar chart showing when they will all be used up in the U.K..*

4. *Draw and describe what was done at Dinorwig in North Wales. Explain why this method of producing energy would not work in the fens. Explain why we might not produce electricity from Dinorwig at night. (3 sides of A4 will be sufficient).*

5. *Explain how a sewage works might be useful in providing future energy. (A brief outline will suffice).*

6. *Draw and describe the principle of how Salter's Ducks work.*

7. *Draw a map of Britain showing where you might place wind generators. Give reasons for your positions.*

8. *On the map show two locations where you might make use of the tide in producing energy.*

9. *Having looked at several different energy sources, suggest which sources we should be looking to in the future. Suggest the part you think nuclear energy will play in producing our energy for the future. (One side of A4 will be sufficient). Your work should be submitted in a file. Later in the year you will be interviewed for an imaginary post at the college as a laboratory technician. It is expected that you will bring this file with you and that you will be able to discuss it's contents. The presentation of material should therefore be of a very high standard. You will be given a submission date for the work.*

Science in the whole curriculum

17 Science and Environment Education

Jackie Hardie and Monica Hale

1 Environmental Education in the Curriculum

The world is undergoing great and complex changes which affect both the environment in which we live and the quality of life. As a consequence, interest in and concern for the environment at local, national and international levels has increased in recent years, as more and more people have realised how fragile are the conditions for the survival of many of the life-forms on our planet.

The future of humankind, and of many other species of life on earth, will depend on the environmental skills that pupils develop and the informed concern that is fostered. They need to be made more aware of their responsibility to care for the environment and of how their decisions and level of participation will ultimately influence the earth's future. Environmental education is an essential component of the curriculum if pupils are to become aware of the factors and processes which interact to shape the environment in both the short and the long term, and to be rational about the problems and solutions. Equally, it helps them to develop a sense of 'place' and 'belonging' which enhances self-esteem, responsibility and the ability to act positively.

The curriculum aims of environmental education suggested by the National Curriculum Council (NCC 1990b p3) are:

1. *to provide opportunities to acquire the knowledge, values, attitudes, commitment and skills needed to protect and improve the environment;*
2. *to encourage pupils to examine and interpret the environment from a variety of perspectives – physical, geographical, biological, sociological, economic, political, technological, historical, aesthetic, ethical and spiritual;*
3. *to arouse pupils' awareness and curiosity about the environment and encourage active participation in resolving environmental problems.*

Environmental education involves *three* elements:

1. education *about* the environment, (in order to develop knowledge and understanding, values and attitudes *in relation* to the environment);
2. education *in* or *through* the environment, (here the environment is used as a resource for developing knowledge, understanding and skills).
3. education *for* the environment, (the exploration of an individual's response to the environment and environmental issues to *ensure the sustainable and caring use of the environment* now and in the future, ultimately leading towards a sense of personal ownership, involvement and care).

All subjects can contribute to environmental education: it is cross curricular in every sense. It allows teachers to extend their work beyond the confines of the classroom, into the immediate environment of pupils, moving gradually from the home and school to the natural and built environments. It provides opportunities for learning at first hand at every Key Stage, using the natural curiosity of pupils who want to learn about themselves and the world around them, so making the world for them a more relevant and understandable place.

2 Science and Environmental Education

Science undoubtedly has a very important contribution to make to environmental education. In *"Opening Doors for Science"*, (ASE 1990) the relationship is stated as:

> *helping pupils to develop and apply science knowledge and skills to make decisions in order to prevent or solve problems concerned with caring for the whole environment. This should be based on their own and other people's evidence (p 5).*

The National Curriculum Programmes of Study (PoS) provide many potential contexts for the development of environmental education. For example at KS4 we have the following:

> *...pupils should be given opportunities to develop awareness of science in everyday life... They should consider the effect of scientific and technological developments including the use of information and control technology on individuals, communities and environments. Through this study, they should begin to understand the*

> *power and the limitations of science in solving industrial, social and environmental problems and recognise competing priorities. (General introduction KS4)*

and again in paragraph (iii) from the PoS for Sc2 "Life and Living Processes", which relates to strand (iii) "Populations and human influences within ecosystems":

> *Pupils should make a more detailed and quantitative study of a habitat... They should explore factors affecting population size,... They should have opportunities, through fieldwork and other investigations to consider current concerns about human activity.. They should relate the environmental impact of human activity to the size of population, economic factors and industrial requirements. The work should encourage pupils to use their scientific knowledge, weigh evidence and form balanced judgments about some of the major environmental issues facing society.*

Similarly there are environmental links in Sc3 strand (iv) "The earth and its atmosphere,"

> *..Pupils should study, through laboratory and fieldwork, the evidence which reveals the mode of formation and later deformation of rocks, and the sources of energy that drive such processes. Pupils should study the scientific processes involved in the weathering of rocks, transport of sediments and soil formation. ...*

and Sc4, as indicated in strand (ii) "Energy resources and energy transfer"

> *...They should be introduced to the ways electricity is generated in power stations from a range of resources, both renewable and non-renewable. By analysis of data, pupils should understand that some energy resources are limited and consider the longer term implications of the world-wide paterns (sic) of distribution and use of energy resources including the "greenhouse effect". They should be given opportunities to discuss how society makes decisions about energy resources.*

Science has much to contribute to the environmental education curriculum, especially in terms of contexts, concepts and ideas. It can, in turn, draw strength from it, as part of an approach to education which is concerned to foster the skills, knowledge and values necessary for pupils to make sense of the world and to contribute constructively to a future characterised by uncertainty and change.

Opportunities for environmental education exist in all other subject areas, in extra-curricular and community based activities and at all levels of schooling. (See NCC 1990b and ASE 1990 where these issues are explored in more detail). So it is necessary to determine:

- *where* learning takes place
- *what* content is taught
- *how* it is taught

3 Issues

3.1 Whole school and departmental approaches

To decide on and write a policy for environmental education, each school needs to reflect on its own unique set of circumstances, but, as pupils have a certain curriculum entitlement and there is a growing consensus about the nature of environmental education, it is likely that school policies on it will have much common ground. As individuals who participate in policy making and curriculum design are far more likely to become actively involved in implementing them, a policy is best planned collectively, so that the curriculum reflects the aims of the school, meets the needs of individual pupils and is taught as a cohesive whole.

A good policy on environmental education should encourage teaching which responds to the curiosity, enthusiasm and current interests and needs of pupils. It should provide for a framework which supports and gives shape and coherence to curriculum planning, whilst allowing scope for diversity, extension and individuality of learning in the classroom.

Environmental education is one of the five cross-curricular themes identified by NCC. All share the following characteristics:

> ...the ability to foster discussion of questions of value and belief; they add to knowledge and understanding and they rely on practical activities, decision making and the inter-relationship of the individual in the community. (NCC 1990a, p3)

It is apparent therefore that the implementation of any one of these themes is a whole school issue. There will also be factors external to the school that may influence the development of policies, such as the DFE, local funding, research findings and the reports of school inspections.

3.2 Planning and managing the environmental education curriculum

The links between environmental education and a number of subject disciplines are both its strength and weakness. For pupils to have a coherent learning experience which will enable them to make progress, careful planning is essential. Subject based and topic based approaches can contribute to achieving the aims in section 1. However HMI and others have indicated that for successful implementation, one person must be given overall responsibility to co-ordinate environmental education and ideally this should be a senior member of staff. The identity of this person and their responsibilities and duties must be made known to all staff in the school.

Successful implementation requires a clear strategy. Fig 1 shows the sequence most schools are likely to follow. However the detail will depend on the internal organisation, management structure, relationship with the governing body and status of the school and their stage of development.

3.3 Questions to determine action

When planning and implementing policies, it can be very helpful to have checklists, to facilitate structured discussion and to make sure important issues are not overlooked. The questions in Table 1 may provide useful indicators of:

- a school's attitude towards environmental education
- areas for individual, departmental or whole school action

HMI (HMI 1989 para 13) have produced a list of questions that should help in the detailed planning and organisation of schemes of work for environmental education whether it be within a subject or as part of a topic or theme. Table 2 (page 316) is a modified version of these.

3.4 Site selection

3.4.1 The school site

The most relevant and easy to use site on which to focus environmental work is the school itself. At their best, school sites have a variety of interesting areas but at their worst they are dull expanses of asphalt and concrete often with buildings in poor repair. Nevertheless every site has the potential for environmental work. Teachers need to be sensitive to pupils' feelings about

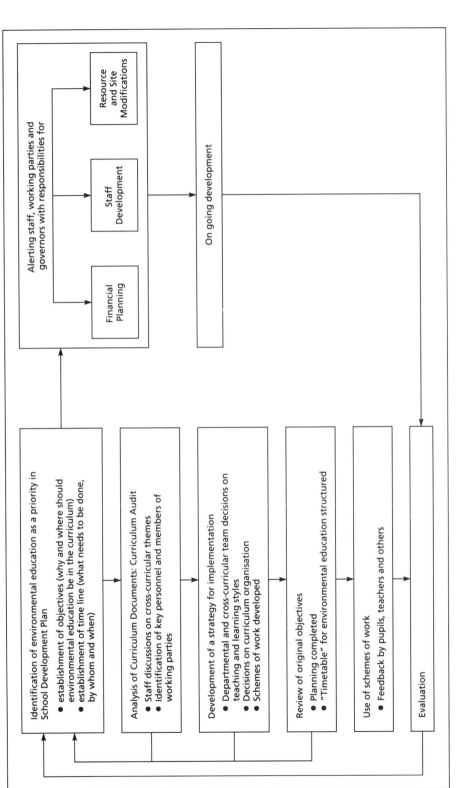

Figure 1 A strategy for environmental education implementation

Table 1 Environmental Education Checklist

The documents on environmental education
- Is there a school policy or guideline on environmental education? If not, is there an action plan for development?
- Is the action plan linked to other aspects of the school plan eg. staff development, financial planning?
- If there is documentation does it need to be updated, reviewed or evaluated? Does the school handbook or prospectus outline the school's attitudes towards environmental care, including the policy on litter, smoking, recycling, energy efficiency, purchasing goods and the teaching of controversial issues?

The management of environmental education
- Has the school ever established environmental education as a priority? If so, when?...and what happened?
- Have there been discussions about how to organise environmental education amongst the staff? If not, why not?
- Is there one teacher with overall responsibility for environmental education? Is there a team of teachers with a range of subject expertise involved in working parties and with planning and teaching?
- How are the discussions and decisions of working groups communicated to senior management and the full staff?
- Has there been a curriculum audit?
- Are those teachers with responsibility for other areas of the curriculum asked to state how far the objectives and content of their schemes/programmes of work contribute, or might contribute, to developing pupils' environmental understanding?
- Do teachers review the activities undertaken in different year groups to ensure coherent curricular progression?

Integration into the curriculum
- Do departments meet to discuss curricular issues? How often?
- Is there a timetabling strategy or the possibility for the occasional suspension of the timetable to encourage and emphasise links between subjects?
- Do existing schemes of work incorporate environmental education?
- Is diversity of opinion amongst pupils encouraged?

Relevance to pupils
Has a curriculum audit revealed if pupils:
- learn about the same environmental topic in different lessons?
- are alerted to major environmental issues, and/or local environmental issues?
- are encouraged to become members of "environmental" or other groups such as R.S.P.B., National Trust, Conservation Volunteers?

Assessment
- What are the arrangements for assessment of progress and achievement in environmental education?
- How do you check on the development of awareness, skills, understanding and values in pupils?
- How is progress recorded? ...and by whom?
- Are achievements in environmental work rewarded? If so, how?

The school building and site
- Does the appearance of the school building, the site, classrooms and corridors indicate an interest in and understanding of the immediate surroundings?
- Are pupils involved in any aspect of planning and maintenance of the environment?

Using first hand experience

(1) The school site

- Is there evidence that the school site is used as a learning resource?
- What opportunities does the school site offer for formal and informal learning such as enjoyment, recreation and play?
- Are the grounds used for extra curricular activities or any curriculum work?
- Is vandalism a problem? If so, has its avoidance or reduction been considered?
- Is a variety of equipment and resources placed around the site to draw attention to changes in the environment? eg weather vane.
- Is the site managed to encourage a diversity of wildlife? Are there habitat areas, ponds, gardens, trees, bird feeding areas? If not, is the development of any such areas possible?
- Are there any plans for enhancing provision on site?
- Is any such plan linked to the School Development Plan?
- Has funding for improvements been sought from any organisations such as English Nature, Princes' Trust, local industry?

(2) Off site

- Do the school staff organise fieldwork in the immediate vicinity or further afield?
- Are such visits designed to facilitate a progression in the development of pupils' knowledge and skills?
- Are visits structured so that they are an integral part of the environmental education programme?
- How is participation in fieldwork recorded?
- Are any staff trained in fieldwork leadership?
- Are all staff aware of the legal requirements of site visits (eg. adult to pupil ratio, charges to pupils, access to site).
- Are school governors aware of any programmes involving off-site visits?

Leading by example: Do we practice what we preach?

- Is the school "litter free"?
- Are litter bins sited where they are needed?
- Does the school recycle anything? eg. re-use envelopes, collect cans. Is the school energy efficient or are there lights left on, windows and doors left open when heating is on?
- What form of heating is used? oil, electricity, gas?
- Does the school avoid the use of aerosols, eg. in art and technology lessons and do pupils appreciate the reasons?
- Does the purchasing policy adopt environmentally friendly products, eg. non-chlorine bleached paper products.

Table 2 Environmental Education: Schemes of work checklist

- Has the topic or issue been tackled before, in some form or other, by the pupils, and if so how does the work build on what has gone before?
- Is it relevant, of interest and value for the pupils?
- What are the main objectives in studying the topic or issue?
- What are the subject areas, programmes of study and attainment targets with which it matches?
- How does the study further develop pupils' understanding, skills and ability to conduct an enquiry and draw warrantable conclusions from it?

- What is the balance of direct and indirect experience and of ideas, skills and areas of knowledge?
- Are there differentiated activities and experiences for particular pupils?
- Are there points at which pupils can develop their own lines of enquiry or areas of interest?
- Are the teaching and learning approaches appropriate for the aims of environmental education?
- What are the resources needed?
- How is the work of each pupil to be recorded and what use can be made of these records for assessment and profiling?
- How is the learning undertaken by the pupils to be evaluated?

the quality of their environment and aware of the consequences of any judgement made about the quality of the local surroundings. A useful national source of help is the *"Learning Through Landscapes Trust"* (LTL) and schools can become members at a small cost. The trust exists to promote improvements to the quality of school grounds and encourage the widest possible educational use of land around all educational establishments. Their research has shown how school grounds can be used as an educational resource and they produce many publications dealing with education about, and in, the environment.

Science departments can exploit the immediate environment and create activity trails and areas which relate specifically to the science component of the curriculum. Suggestions for such trails can be found in School Science Reviews (See for example Borrows 1984, Forster 1989).

3.4.2 Off-site

Local Planning and Leisure departments will have details of sites which can be visited by schools and information about local environmental groups. These are a source of useful, relevant, topical ideas and speakers – expert and lay – who may be willing to visit a school.

Some off-site visits may involve a charge and teachers should know their own school's charging policy. These policies resulted from legislation in the Education Reform Act 1988, and are particularly significant for educational visits during school hours or in connection with the National Curriculum and public examinations. Reductions in education budgets may jeopardise the future of off-site visits, so careful preparation and an analysis of the benefits, both social and educational, gained by pupils may be needed to justify their continuation.

The senior management of the school is responsible for ensuring that leaders of school parties and accompanying adults have the necessary personal qualities, experience and training and that any visit or journey complies with the regulations of the LEA or, in the case of independent schools, the governors. The staff/pupil ratio will vary according to the nature of the activities and the age of the pupils, and whether the party is single sex or mixed. The LEA or Governing Body will have regulations about the requisite ratios.

If the off-site visit is to a residential centre, such as one run by the Field Studies Council or the Youth Hostels Association (YHA), parents must be informed. They should be told the purpose of the trip and given as many details of the itinerary as possible.

3.5 Teaching and learning styles

Environmental education requires an approach to teaching and learning which aims to develop pupils' ability to:

- work cooperatively
- communicate their findings and views to others
- be tolerant of the opinions of others
- make decisions
- act in a responsible and caring way for the environment

A variety of experiences and learning contexts should be used, as well as flexible approaches to learning to suit particular needs, abilities and interests. Some suggestions are given in table 3.

An example of the application of these ideas is shown in table 4.

Table 3 Learning experiences

Type of experience	Comment
participatory and cooperative learning in small groups	encourages respect for others' views, challenging of facts and their interpretation, and team working
pupils taking responsibility for their own learning and assessment	promotes thinking and action skills
first hand experiences, to enjoy and explore the environment and take positive action in caring for it	effectiveness enhanced if time is set aside for pupil discussion of their views and beliefs, and also of current issues
problem solving and enquiry based learning	
working with and learning from people in the local community directly involved with the environment	e.g. environmental health officers, parks staff, planners, architects, wildlife groups, industrialists, politicians: such work enhances motivation and interest and helps to form values and attitudes.
use of secondary sources of information	provide additional evidence and can further stimulate interest,
use of videos, games, drama and simulations	particularly on more global matters.

Table 4

Example: Improving the local area
Source: Non-statutory Guidance for Geography (NCC 1991)

National Curriculum cover: sections of PoS for Geography, Science, English, Mathematics, Technology: also parts of cross-curricular themes
Key stages could use at: all
Types of site: school playgrounds, playing fields, local open spaces etc.
Content: Pupils design a landscape plan in which they:
(a) extend their communication skills by:
    devising questionnaires; writing reports to be presented to profession-als; interviewing local residents, teachers, governors, people who work at the school and groups who use the school premises out of school hours; and negotiating with council officers, teachers and other professionals;
(b) design a site which may have an improved appearance and is *safe; is a *conservation area; which could be used as a *resource for field-work and *may be a facility for the community;
(c) carry out and evaluate work on the site.

Examples of science links
Key stage 3: Sc2: Programme of Study
Strand (iii) *"Pupils should study a variety of habitats at first hand"* ... eg. the local site, which could be a pond, woodland, or the school paths and play areas.
    Strand (iv) ..."*they should investigate ways of improving the local environment"* eg by planning a habitat area and considering the organ-isms to be introduced so as to facilitate its use either as conservation or fieldwork resource, perhaps by:

1) planting nettles (food plants for caterpillars), Buddleia and other plants that attract butterflies.
2) growing wild flowers from seed and then transplanting seedlings into a prospective meadow.
3) creating tracks through the area.

(*Links with science can be developed through these aspects.)

Some of the issues that will be tackled in environmental education will be sensitive and controversial. HMI, (HMI 1989) advise that:

> *...teachers should not preach or condemn; their task is to explore ideas with pupils and help them become better informed. The purpose of discussing controversial issues cannot be to give young people a complete understanding or knowledge of them; no one has this. But misunderstanding and distortion can be lessened through the provision of well-founded information, and ill-informed value judgements can be avoided by giving pupils practice in considering the messages bombarding them from the various groups interested in the matters concerned. In general, if teachers are asked for their own opinions, it seems sensible that they should give them, while at the same time making it clear that other reasonable and serious people, including the pupils' parents and guardians may legitimately hold different views.*

Some controversial environmental issues are tackled in SATIS materials, and these contain advice on teaching strategies. (See chapter 20)

4 Case Studies

At KS3 environmental education may be pursued through the study of separate subjects, but the links between them and environmental issues must be planned, and pupils must have access to first hand experiences in the environment. An alternative strategy is to design integrated courses which focus on the environment as one of a number of important themes or areas of study, using topics such as 'The school site' or 'Forests and paper.' Some schools may decide to deal with all cross-curricular themes in this way.

At KS4, the developments in GCSE and the National Curriculum are such that pupils have limited options. All will follow science courses, 'single', 'double' or three separate sciences:. Geography, History and Technology will exist in a variety of formats. Such variation could have made it difficult to ensure all pupils have access to the environmental education to which they are entitled, but the documentation produced for GCSE Science syllabuses reveals that boards have taken into account Resolution 88/C 177/03 (24.5.88) of the Council of Ministers and have incorporated environmental issues into their syllabuses. Examples are:

Co-ordinated Science (The Suffolk Development, syllabus 1777 MEG),
    Units B2, B6, C6, C7, P2
ULEAC *Science Syllabuses* 1531 and 1538.

Those senior teachers (see section 3.2) responsible for coordination, planning and curriculum audit should compare schemes of work developed from these syllabuses, and ensure that environmental education is an explicit component of the curriculum and pupils are given coherent learning experiences.

In a secondary school, timetabling arrangements should be flexible so that occasionally pupils may work with a number of teachers over a period of time on an environmental theme, for example, pollution, drought, famine, or energy. One idea is to organise a "Green Week" where the whole school may focus on environmental topics or devote the time to fieldwork. In this way the different specialisms of teachers can complement each other.

Planning for coherence and continuity in learning experiences and for progression in pupils' development is one of the most difficult tasks in environmental education. The NCC booklet, *Curriculum Guidance 7* (NCC 1990b) provides case studies for key stages 3 and 4 to illustrate the knowledge, skills and attitudes which are relevant to the particular activity being described (see particularly pages 28-37).

Schemes of work
There is plenty of advice available on the development of schemes of work. One good source is the *Non-statutory Guidance* (NSG) for geography already mentioned. (NCC 1991)

Another example from this source is the study of a locality such as the sea shore or coast: although it was originally developed for KS2 it has wider applicability. The work could be done as part of an off-site project, perhaps when the timetable is suspended. This is shown in Table 5. Again, points linking the study to science are asterisked.

Threads of environmental education can span across or link together a variety of areas of experience. Table 6 shows two possible approaches to "Energy" : approach 'A' probably requires a greater time commitment than 'B'.

Here the links have been established by content. Teachers must also consider the teaching styles and resources to be used so that a coherent programme is presented to pupils.

Table 5 The study of an off-site locality

Key Questions	Learning Objectives	Pupil Activities
	Concepts	*In the field:*
What are the similarities and differences between the local area and visited locality?	Changing landscape Land-use The protection of the environment	group/individual observation, sketching 1* Observation of flora, fauna and rocks Beach transect
	Skills	
How is tourism important to that locality	2* Enquiry skills – data collection, observation, investigation Map work Atlas work Use of primary and secondary sources	Group work to measure, record, sketch, photograph Use of compass Observation and sketching of a landscape Discussion of land-use Orientate OS map on site; on beach; at viewpoint; at monument, eg. castle, keep
	Content	
	3* Wearing away of sea coast Depositing the eroded material Protection of coasts Use of coastal areas for leisure activities, hotels, retirement homes	Draw sketch maps Draw cliff face, sketch and label Record temperature each day of visit to compare with other simultaneous records at school.

Notes on Table 5

1* overlap with KS3, Sc2, PoS strand (iii)

> *Pupils should study a variety of habitats at first hand and .... investigate the range of seasonal and daily variation in physical factors, and the features of organisms which enable them to survive these changes.*

2* overlap with the development of a problem solving strategy (Sc1) eg.
    (a) how are seaweeds distributed on the shore?
    (b) how does the temperature of a rock pool change during tidal exposure?

3* overlap with KS3, Sc3, PoS, strand (iv)

> *Pupils should investigate, by observation, experiment and field work, the properties and formation of... rocks and link these to major features and changes on the earth's surface*

Table 6 Example: Energy

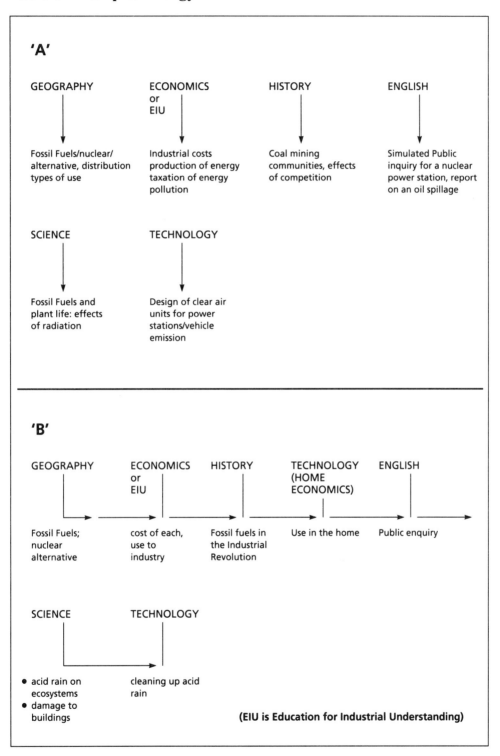

'A'

GEOGRAPHY	ECONOMICS or EIU	HISTORY	ENGLISH
Fossil Fuels/nuclear/ alternative, distribution types of use	Industrial costs production of energy taxation of energy pollution	Coal mining communities, effects of competition	Simulated Public inquiry for a nuclear power station, report on an oil spillage

SCIENCE	TECHNOLOGY
Fossil Fuels and plant life: effects of radiation	Design of clear air units for power stations/vehicle emission

'B'

GEOGRAPHY	ECONOMICS or EIU	HISTORY	TECHNOLOGY (HOME ECONOMICS)	ENGLISH
Fossil Fuels; nuclear alternative	cost of each, use to industry	Fossil fuels in the Industrial Revolution	Use in the home	Public enquiry

SCIENCE	TECHNOLOGY
• acid rain on ecosystems • damage to buildings	cleaning up acid rain

(EIU is Education for Industrial Understanding)

Jackie Hardie taught science in East London and is now Science Adviser in Enfield. She has written or edited many science texts, been on the editorial board of The Journal of Biological Education, and served on the NSAIG of ASE and the Royal Society Joint Biology committee.

Monica Hale is Senior Education Officer at the Council for Environmental Education. She was Project Director of the Urban Spaces Scheme, after lecturing in biogeography, researching in heathland ecology and pedology and working in industry.

Bibliography

Available from ASE Bookshop

Learning Through Landscapes Trust publications:

Booth P R (1990) *Ecology in the national curriculum – A practical guide to using school grounds*, LTL.

LTL (1991) Making the best of your school grounds (Video), LTL.

RSPB/LTL (1991) Wildlife and the school environment, RSPB/LTL.

Young K (1990) *Using school grounds as an educational resource*, LTL.

ASE publications:

ASE (1990) *Opening doors for science*, ASE.

ASE (1986-1992) *Science and technology in society (SATIS) series*, ASE.

NCC Curriculum Guidance Series

NCC (1989a) *A framework for the primary curriculum*, NCC.

NCC (1989b) *A curriculum for all*, NCC.

NCC (1990a) *The whole curriculum*, NCC.

NCC (1990b) *Environmental education*, NCC.

Borrows P (1984) *The Pimlico chemistry trail*, School Science Review, vol 66, no 235, pages 221-233.

DES (1989) *From policy into practice*, DES/HMSO.

DES (1990) *The outdoor classroom*, DES/HMSO.

Forster S (1989) *Streetwise physics*, School Science Review, vol 71, no 254, pages 15-22.

HMI (1989) *Environmental education from 5-16* (Curriculum Matters series, no 13), DES/HMSO.

INSET for Environmental Education 5-16: Introductory Activities, (1992) Longman.

Jennings, Arthur (1986) *Science in the locality,* Cambridge University Press.

MEG (1992) *Co-ordinated science (The Suffolk development) syllabus 1777.*

NAEE (1992) *Environment audit (towards the school policy for environmental education),* NAEE.

NCC (1991) *Non-statutory guidance for Geography,* DES/HMSO.

Williams J and Dowdeswell W H (1990) *Ecology for the National Curriculum: Investigations in the school curriculum,* Unwin and Hyman.

Useful Addresses

British Trust for Conservation Volunteers
36 St Mary's Street, Wallingford, Oxon, OX10 0EN. Tel: 0491 39766

Council for Environmental Education, The University of Reading, London Road, Reading, Berks. Tel: 0734 756061

Council for the Protection of Rural England, Warwick House, 25 Buckingham Palace Road, London SW1W 0PP. Tel 071 976 6433

English Nature, Northminster House, Peterborough PE1 1UA Tel: 0733 340345

Field Studies Council, Preston Montford, Montford Bridge, Shrewsbury, SY4 1HW. Tel: 0743 850674

Friends of the Earth, 26-28 Underwood St , London N1 7JQ. 071 490 1555

Learning Through Landscapes Trust, Third Floor, Southside Offices, The Law Courts, Winchester SO23 9DL. Tel: 0962 846258

National Association for Environmental Education, Wolverhampton Polytechnic, Walsall Campus, Gorway, Walsall WS1 3BD. Tel: 0922 31200

National Trust, 36 Queen Anne's Gate, London SW1H 9AS. Tel: 071 222 9251

Princes Trust, 8 Bedford Row, London WC1R 4BA. Tel: 071 430 0524

Rainforest Action Group, 9 Cazenove Road, London N16 6PA.

Royal Society for Nature Conservation, 22 The Green, Nettleham, Lincoln, LN2 2NR. Tel: 0522 544400

Royal Society for the Protection of Birds, The Lodge, Sandy, Beds ST19 2DL. Tel: 07676 80551

Survival International, 310 Edgware Road, London W2 1DY. Tel: 071

18 Science and Health Education

Angela Dixon

1 Introduction

Health Education, especially Sex Education, has for a long time been a fully integrated component of the biology curriculum. With the Education Reform Act of 1988, Health Education became incorporated by Act of Parliament into the curriculum of all pupils at state maintained schools up to age 16. It is embedded in the National Curriculum for science more specifically in Attainment Target 2 (Sc2), *Life and Living Processes*. It also exists as a separate cross-curricular theme.

Unless care is taken to prevent it, there is a danger, linked partly to this double exposure and to the strong historical links with biology, that some aspects of health education will receive more attention and emphasis than others. Furthermore, more recently developed components may get omitted altogether. It is for these reasons amongst others that the National Curriculum Council in its Curriculum Guidance 5 – *Health Education* (NCC 1990) advocates the establishment of a whole school policy for Health Education.

2 Whole School Policy for Health Education

While it is important to have a school policy covering the whole area of Health Education, it is *essential* for there to be a school policy on Sex Education. This has already been singled out for special treatment by the government and so it seems appropriate to deal briefly with this special case before looking at more general issues of Health Education policy. It is discussed in more detail on page 332.

Sex Education is already covered by the Education (No 2) Act of 1986 which *requires* school governors to decide whether sex education shall be a curriculum component and if so, what should be taught. They are required to have a policy on it even if this states that there isn't one!. DES Circular 11/87 gives further guidance on sex education in the curriculum.

Where the governors have a policy in support of Sex Education they are *required* to have a curriculum for it with which teachers *must* comply.

Generally speaking, such a curriculum is drawn up by the school at the request of the governors and endorsed by them.

Since Sc2 *(Life and Living Processes)* includes the human life cycle and since this Attainment Target is, as a result of the 1988 Education Reform Act, a legal requirement, it is tempting to think that the earlier 1986 Act has been subsumed in the latter. While this appears to be the case, Sex Education is, properly, seen as more than the human life cycle, and whereas governors cannot prevent the teaching of topics contained in the National Curriculum, they may, legally, *direct* teachers how and what to teach on, for example, contraception, abortion and safer sex in connection with HIV transmission. It is still technically and legally possible for school governors to proscribe the teaching of Sex Education at the same time as the staff are, by law, teaching about the human reproductive cycle and the transmission of HIV.

2.1 Health Education Policy

In the view of the National Curriculum Council (Curriculum Guidance No.5 1990) a school's policy for Health Education should cover four main areas:

(i)   content – knowledge and understanding;
(ii) skills and attitudes;
(iii) management
(iv) progression and continuity.

2.1.1 Knowledge and Understanding

It is important – some would argue essential – that pupils are given, or helped to acquire, correct factual information on which to base decisions affecting their future lifestyle. Pupils should also understand how the knowledge they possess can be used to promote a healthy (or healthier) lifestyle or to prevent illness or disease. The World Health Organisation's definition of health as a *human right which is not just the absence of disease but a state of total physical, psychological and social well-being* is usually accepted as the aim of all Health Education.

2.1.2 Skills and Attitudes

It follows from the WHO definition of health that pupils should be given opportunities to explore their own attitudes to the information they acquire and, in addition, to practise the interpersonal skills that are required for them to confirm these attitudes and beliefs and/or to alter and modify them over time. Whereas much of the knowledge and understanding component of Health Education may occur primarily in science lessons, the attitudes and skills component is – or was – more likely to be found in Personal and Social Education (PSE) and/or in tutorial time. There is however an increasing blurring of these boundaries. Science has for some time acknowledged that attitudes to things scientific, and the ability to debate and argue about their relevance to and influence on daily life-styles, are essential to a proper understanding of science. PSE has likewise accepted that attitudes spring from knowledge and understanding and that, without information, arguments quickly become circular and sterile. It is important therefore that any Health Education Policy includes a management component.

2.1.3 Management

Management of a school's health education policy would, in addition to agreeing the knowledge, understanding, attitudes and skills components, also involve decisions about the most appropriate location for these. So it would entail apportioning responsibility amongst the subject disciplines and any cross-curricular areas such as P.S.E. This is necessary not only to avoid unnecessary repetition of material (not contraception / AIDS / the environment again!) but also to coordinate approaches, teaching styles and resources and to match them to the teachers and the topics. Many teachers have considerable expertise in some aspects of health education, gained at some expense not only to themselves but also to the school. It makes sound financial sense in today's climate of LMS for that expertise to be maximised. Rather than have several teachers teaching their variation of the same topic in their respective subject curricula, it may be more cost-efficient to use the same teacher several times. This would also help to reinforce the transferability of content for pupils. Teachers with special expertise and qualifications could also be involved in school-based INSET, enabling colleagues to acquire additional knowledge and skills.

2.1.4 Pupils: continuity, progression and differentiation

It is a truism that pupils progress at different rates throughout their schooling, and while the National Curriculum has set particular Attainment Targets for the different Key Stages, only part of the Health Education curriculum is included in the Core and Foundation subjects. Although cross-curricular work requires subject boundaries to loosen and allow permeation, it is of little help to the pupil if the structure of continuity, progression and differentiation also disappears. Generally speaking, such structure becomes more important in cross-curricular work and especially when, as in Health Education, it is the individual pupil's attitudes which are the educational target. It may therefore be necessary to repeat factual information more than once throughout the key stages but it must be remembered that although the facts themselves will not be different the context will be, and it is this context that holds the key to facilitating attitude clarification, confirmation and where appropriate, change. As pupils get more mature they will be able to accept the challenges of more complex issues and this should be reflected in the overall policy for Health Education.

2.2 Health Education in Context

There is little point in one department addressing health education issues, if these are unsupported or, even worse, undermined by other discipline areas. Attitudes are very susceptible to both positive and negative images – (hence the power of advertising). Environmental awareness, for example, is not restricted to the area outside school. Litter from a school tuck shop is a very powerful health education message, as is a well trodden footpath across a flower bed. Likewise the safety training that goes on in the science department can be undermined by lack of care in other areas.

Since health education is concerned with lifestyle, it will spread beyond the school grounds into the pupils' homes and surrounding area. This should encourage the school to include people from the local community in its health education programme. Local shopkeepers, for example, would be able to provide very important information about legislation covering the storage of foodstuffs, especially perishable goods. Builders and electricians are good sources of safety regulations, and the local planning office has plenty of examples of what is and is not considered desirable in buildings. It is essential that pupils appreciate that the health education they receive in school has its counterpart in the life and work of adults both within the school and in the community outside.

3 Components for a Health Education Curriculum 5–16

Curriculum Guidance 5 – *Health Education* – (NCC 1990) lists nine components which, in the Council's view, should be built into a Health Education Curriculum. They are:

- substance use and misuse
- sex education
- family life education
- health related exercise
- safety
- food and nutrition
- personal hygiene
- environmental aspects
- psychological aspects

It would be possible, given the time and curriculum flexibility, to incorporate each of these into the science curriculum. However, it is more realistic to focus on those components which are already part of the National Curriculum for science. These appear to be: Substance use and misuse, Sex education, Safety, Food and nutrition and Environmental aspects. The last of these is dealt with in chapter 17. The inclusion of these health education components in the National Curriculum for science does not mean that they do not occur elsewhere. It is the task of the managers of the school's Health Education policy to indicate where and by whom these components will be taught.

There is much to be gained by teaching topics more than once within the same subject area as well as in other subjects, so long as this teaching is coordinated so that both teachers and pupils are aware of its multiple cover. Looking at environmental aspects in both Science and Humanities, for example, should enhance the study of the components in both areas.

4 Case Studies

The remainder of this chapter comprises a more detailed consideration of some of the health education components identified above as most closely relating to the National Curriculum for Science ie:

1 Substance use and misuse
2 Sex education
3 Safety
4 Food and nutrition

4.1 Substance use and misuse

This topic occurs throughout the Key Stages and Programmes of Study. This case study is based on Sc2, KS4 and the corresponding PoS.

pupils should have opportunities to consider the effects of solvents, alcohol, tobacco and other drugs on the way the human body functions.

The use and misuse of tobacco is well established as a topic in teaching about the respiratory system (lungs, heart, bloodstream etc.). Solvent and alcohol use and misuse are more recent additions, but both can be included readily in the teaching of the body systems. Solvent misuse can be incorporated into the teaching of either the respiratory system or the nervous system and there is a strong case to be made for repetition in both topics. Alcohol use and misuse also fit neatly into the teaching of the nervous system, but may be taught equally appropriately in microbiology and/or biotechnology since alcohol is a product of the anaerobic respiration of a micro-organism, yeast.

4.1.1 Knowledge and Understanding

Before pupils can decide whether to use these substances themselves, they need to know the laws governing their sale and consumption. Solvents are sold legally for normal use but become illegal substances when inhaled. Pupils also need to know the legal requirements and medical advice on 'safe' drinking limits, especially when driving, and the current medical view of the effects of alcohol and tobacco consumption.

Tobacco and alcohol are also important for their economic aspects. Both are heavily taxed and as such represent government income: both are part of our social structure. Tobacco is becoming less acceptable socially, but alcohol remains an integral component of our culture; in religious ceremonies, in celebrations, at social gatherings etc. In some circumstances it is safer to drink alcohol than water. There is currently a wealth of material provided by those seeking to moderate the use of potentially harmful substances. Much of this material is available free of charge or at a nominal cost. This is an area in which pupils should be encouraged to check their own knowledge assumptions and find out what they do not know.

4.1.2 Attitudes and Skills

Although, strictly speaking, pupils should be allowed to form their own opinions about use and misuse of substances such as alcohol, it must be acknowledged that Health Education is not neutral as it supports decisions leading to a healthy lifestyle. Furthermore, it is no part of any education to condone, implicitly or explicitly, the breaking of the law by young people. In debating such questions as:

- "Should tobacco/alcohol advertising / sponsorship be illegal?"
- "What would happen if everyone stopped smoking?"
- "How useful is alcohol?"

teachers will need to be prepared to confront their own attitudes and lifestyles. How many of us have driven, suspecting that we are over the limit of alcohol in our bloodstreams? How many of us have "turned a blind eye" to under-age drinkers and smokers? How many pupils are still gently bathed in tobacco smoke each time the staff-room door opens? (The requirement to teach about tobacco use and misuse has been the spur to many smokers to stop smoking.)

There are obvious links here with active tutorial methods and it is important that links with PSE and tutorial work enable not only a free exchange of teaching materials but also styles and strategies. It does not matter if the emphasis is different but it is important that the information transmitted in both areas is substantially the same and correct, and that the aims are congruent.

4.2 Sex Education

4.2.1 Human Reproduction

Human Reproduction is a topic which features at all Key Stages and levels in the National Curriculum.

(Sc2) Key Stage 1 level 3(a) *"know the basic life processes common to humans and other animals"*

(Sc2) Key Stage 3 PoS *"they should study the human life-cycle, including the physical and emotional changes that take place during adolescence, the physical and emotional factors necessary for the well-being of human beings in the early stages of their development, and understand the need to have a responsible attitude to sexual behaviour"*

(Sc2) Key Stage 4 Level 10 (b) *"Understand how DNA replicates and controls protein synthesis by means of a base code"* and (Sc2) Key Stage 4 PoS *"Using the concept of the gene they should explore ..... how sex is determined in human beings and how some diseases can be inherited"*.

4.2.2 *Contraception*

The PoS for Sc2 at KS3 does not prescribe any teaching about contraception. It does, however, include the requirement to teach about the size of populations, including competition for resources, and that pupils "should study the effects of human activity..... on the Earth's surface". Since the most rapidly expanding population of mammals is that of humans, most teachers would find it difficult to teach this whole area without some reference to family planning and thence contraception. In so doing however, they must pay attention to the governors' policy for sex education. Those teaching in Roman Catholic schools will have to teach the topic in accordance with the Church's view on contraception.

4.2.3 *HIV and AIDS*

Both the National Curriculum for Science and Health Education as a cross-curricular theme require pupils to be taught about the Human Immuno-deficiency Virus, and the way its spread is linked to sexual behaviour. The risk from infection by the HIV virus is now as great amongst heterosexuals as homosexuals; this was not the case at the start of the present epidemic.

There is currently some concern amongst more fundamentalist religious groups about the desirability of teaching pupils about the need for safer (i.e. protected) sexual intercourse on the grounds that it introduces and indeed promotes sexual behaviour which, in their view, should be restricted to legally married couples. Such groups are seeking to have the requirement to teach about AIDS removed from the National Curriculum. Sexuality and sexual intercourse, being the emotive issues they are, have previously and will no doubt continue to be surrounded by such controversy. In their time both abortion and the contraceptive pill have been the focus of heated debate, argument and disagreement as curriculum topics. It is important to remember, however, that the Education (No 2) Act 1986 gives governors the right not to allow sex education in their school and it is possible that AIDS education may be added to this.

4.2.4 *Clause 28*

Clause 28 of the Local Government (1986) Act expressly forbids the presenting of homosexual or lesbian partnerships as normal. Many teachers have taken this to mean a ban on any mention of the words homosexual and lesbian (or their more common synonyms) whereas the clause was intended to prevent the presentation of homosexual partnerships as a viable family alternative to the more traditional heterosexual ones.

There is nothing in the Education (No 2) Act to prevent teachers from discussing homosexual and lesbian life styles with their classes so long as this does not lead pupils to presume that social mores and more particularly, the law governing sexual acts, treat heterosexuality or homosexuality as equally acceptable. Furthermore it is implicit in Clause 28 that teachers should not set out to promote homosexuality as an acceptable alternative to heterosexuality.

4.2.5 Keeping a sense of proportion

The following are questions taken from recent press reports worldwide.

What are the pros and cons of abortion; (consider for example the views of the R.C. Church, Pro-Life, and women's rights?)

Should fathers be able to prevent abortion?

Should prostitution be organised?

Who should carry condoms?

That such issues are commonly raised in the press indicates that they are likely to be seen as immediately relevant to young people. Nevertheless it is possible, without proper management, to overdo the topic. The apocryphal story is told of a young male science teacher seeking to motivate a rather apathetic group of Year 11 pupils with the promise of "sex next lesson". On receiving a chorus of groans and "not again" he enquired amongst his colleagues where else the topic appeared. To his dismay (but not a little amusement) he found that only two subjects, Maths and Technology, did not consider Sex Education as a legitimate curriculum component – and they were working on it! This story also underlines the need for a whole-school policy for each component of health education and for individual teachers in the various departments to be aware of what it is and of their own contribution to it.

4.3 Safety

Safety is an issue that permeates science lessons from the first lesson in a laboratory until pupils leave school. In the National Curriculum for science, safe working is an essential constituent of Attainment Target 1 – *Scientific Investigation*.

One of the first things done in science at the start of Year 7 is to draw up a list of rules for the laboratory which incorporate the safe handling of materials and emphasise responsibility for self and for others working in the same room. The Control of Substances Hazardous to Health (COSHH) (see page 140) now requires each institution to undertake a Safety Audit of all materials used on the premises and of course the science labs are included in this. The emphasis of this legislation, together with that of the Health and Safety at Work Act, is to make pupils increasingly responsible for monitoring their own safety and that of others and thus that they should become increasingly independent of adult supervision.

Nevertheless, since teachers are considered to be *in loco parentis,* it is important for them to keep reminding pupils of their safety responsibilities, and of course most of them do, verbally or in writing and frequently both. First Aid is generally recognised as an essential component of safety, and science departments are now required, as part of the Safety Audit mentioned earlier, to show that they have made provision for appropriate First Aid cover. It is no longer acceptable that the school nurse is a qualified first aider and in some local authorities school nurses have been directed not to give it. Acceptable qualifications have to be renewed every three years by an agreed validating body and it is highly desirable that each member of the science staff should be thus qualified. Furthermore, as a consequence of the recommendation (Curriculum Guidance 5, NCC 1990) that First Aid should be taught at Key Stage 2, it is likely that an increasing number of pupils will enter secondary schools with basic experience in it. The science department will need to acknowledge these skills and, ideally, should incorporate and develop them in its own safety procedures.

This is another topic for coordination, since other curriculum areas, notably P.E., and Technology, will have similar needs. Safety is, of course, as much a matter of attitudes and skills as it is of knowledge and understanding. Risk-taking is an adolescent trait, one that is essential in bringing young people into direct contact with new and challenging situations in which they can learn about their strengths and weaknesses and come to terms with their limitations. If the teaching about safety is too repressive or confining, it could result in over-anxious youngsters, too frightened to venture far from home or school and over-dependent on adults. Safety teaching must aim to balance caution with challenge so that risk assessment becomes the first stage of any activity and "I didn't think" is no longer heard as an excuse.

4.4 Food and Nutrition
Knowledge and Understanding

> Requirements of a balanced healthy diet [Sc2]
> Observe, measure and manipulate variables [Sc1]

Attitudes and Skills

> Are all low-fat yoghurts healthy?
> Why do we give chocolates as presents?

More often than not the question of what makes a balanced diet has been taught theoretically within the science curriculum with little or no reference to work done in other areas such as Home Economics (Home Technology) and, except in rare circumstances, irrespective of the School Meals Service and the school tuck shop.

The corresponding guidelines for Health Education in this area at KS3 state that pupils should:

- *know that individual health requires a varied diet;*
- *understand malnutrition and the relationships between diet, health, fitness and circulatory disorders;*
- *understand basic food microbiology, food production and processing techniques.*

This seems an ideal topic in which to demonstrate not only the links to be made between Sc 1 and 2 but also the way Health Education can be used in a cross curricular way to make links with other faculties (Humanities, Design Technology) and non-academic areas of school life (School Meals Service, School Tuck Shop).

Having provided the pupils with the knowledge and understanding they require to make informed judgements, they could be invited to investigate the healthiness of one or more 'health foods' e.g. low fat yoghurts. This would involve them in observing and recording contents, presenting them in an easily comparable way and then evaluating the health food for healthiness (many low fat yoghurts for example contain a great deal of sugar) and cost.

The question of chocolates as presents has links with social history and culture and the whole issue of luxury foods – the "naughty but nice" syndrome. Why do some foods have this special label? An enquiry into the equivalent of chocolates, champagne, caviar etc. in other cultures would make a good cross-cultural investigation.

Investigations into the healthiness or otherwise of purchases from the tuck shop or school canteen could lead to the issue of health education through controlled purchasing.

- Should the school tuck shop be required to stock only 'healthy' foods such as fruit (fresh and dried) nuts, low sugar drinks and snacks?
- Should the School Meals Service respond to market forces and serve chips with everything? How wide a choice should be available?
- What is the acceptable balance between providing what pupils want to eat and will therefore purchase and providing healthy food that is left untouched and is, consequently, wasted?
- Generally speaking prepared 'health foods' are more expensive than more readily available varieties. Is such a price differential acceptable?
- To what extent is the hypothesis that behaviour in children is a result of what they eat tenable?
- Were we really better (i.e. more healthily) fed during the 1939-45 war with food rationing?

5 Summary

There is already in the science National Curriculum a significant amount of Health Education. If Health Education is to achieve its cross curricular status, it is essential that all curriculum areas co-operate. Now that all pupils are required to study a balanced science curriculum to 16+ the science faculty is well placed to contribute to, and perhaps even to initiate, cross-curricular studies. Science teachers have, in recent times, demonstrated their willingness to teach across the traditional science specialisms and to create cross-disciplinary science teaching schemes where these appear to be in the pupils' best interest. They are thus well placed to make a significant contribution to Health Education both from within the National Science Curriculum and also in cross curricular initiatives.

Angela Dixon taught in secondary and primary schools before joining Bristol University where she works with pre-service and in-service science teachers. She is a former Chair of the Association for Science Education and a Fellow of the Institute of Biology.

Bibliography

British Council (1991) *Health Across the Curriculum* Longman

Clarity Collective (1990) *Taught not Caught* Clarity Collective

D.E.S. (1991) *HIV and AIDS. A Guide for the education service* D.E.S.

Ewles L. and Simnett I. (1991) *Promoting Health. A practical Guide to Health Education* John Wiley & Sons

Hawes H. (1988) *Child-to-Child: Another Path to Learning* UNESCO

Health Education Council (1984) *Smoking and Me – A teachers's guide* Health Education Council

Health Education Council (1983) *Health Education 13 – 18* Forbes Publications

Health Education Authority (1988) *Teaching about HIV and AIDS* H.E.A.

Health Education Authority (1991) *Health Authorities and Health Education* H.E.A.

Heathcote G. et.al (1989) *In-Service Provision for Teachers of Health Education* H.E.A.

Institute of Biology (1989) *AIDS: A Challenge in Education* I.O.B.

Kelly P.J. and Lewis J.L. (1987) *Education and Health* Pergamon Press

National Curriculum Council (1990) *Curriculum Guidance No.3 The Whole Curriculum* NCC

National Curriculum Council (1990) *Curriculum Guidance No.5 Health Education* NCC

Perigo B. (1989) *The National Curriculum and Health Education* Oxford County Council

Pring R. (1984) *Education as Health Promotion: an Educational Perspective* in Campbell (ed) (1987) *Health Education Youth and Community.* Falmer Press

Tilford S. et al (1990) *Health Education* Chapman and Hall

Went D. (1985) *Sex Education. Some Guidelines for teachers* Bell and Hyman

19 Economic and Industrial Understanding (EIU) in Science Education

David Sang

EIU is one of the cross-curricular themes of the National Curriculum. This section looks at the meaning of EIU, how we can set about extending its influence in our teaching, and what returns we can hope to get for our effort. Industry in this context is taken in its broadest sense, to include large and small scale industry, business, commerce, national and local government, public and private sectors; anything which contributes to the economic activity of our society.

1 Science, Industry and our Pupils

Science and industry are intimately linked. We all live in an advanced industrial society and consume industrial products: many of our pupils will go on to work in industry. Science forms the basis of many technological products and processes and can help us interpret and understand the impact of industry on the environment and the way we use industry's products in our lives.

We can often help pupils to see science as more attractive by setting it in an economic and industrial context. Instead of discussing the deformation of a rectangular beam under stress, we can model crumple zones in cars: instead of discussing the importance of conserving water resources, we can visit a sewage works.

2 Knowledge, Skills and Attitudes

The NCC has produced several publications outlining the knowledge, skills and attitudes which might be developed within EIU, some more relevant to science education and a particular Key Stage than others.

Knowledge and Understanding of:

- *key economic concepts, such as production, distribution, and supply and demand;*
- *how business enterprise creates wealth for individuals and the community;*

- *the organisation of industry and industrial relations;*
- *what it means to be a consumer, how consumer decisions are made and the implications of these decisions;*
- *the relationship between economy and society in different economic systems;*
- *technological developments and their impact on life styles and work places;*
- *the role of government and international organisations (for example, the European Community) in regulating the economy and providing services.*

Analytical, personal, and social skills, including the ability to:

- *collect, analyze and interpret economic and industrial data;*
- *think carefully about different ways of solving economic problems and making economic decisions;*
- *distinguish between statements of fact and value in economic situations;*
- *communicate economic ideas accurately and clearly establish working relationships with adults outside school;*
- *co-operate as part of a team in enterprise activities;*
- *lead and take the initiative;*
- *handle differences of economic interest and opinion in a group;*
- *communicate effectively and listen to the views of others on economic and industrial issues.*

Attitudes, including:

- *an interest in economic and industrial affairs;*
- *respect for evidence and rational argument in economic contexts;*
- *concern for the use of scarce resources;*
- *a sense of responsibility for the consequences of their own economic actions, as individuals and members of groups;*
- *sensitivity to the effects of economic choices on the environment;*
- *concern for human rights, as these are affected by economic decisions.*
  (NCC 1992)

3 Putting it into Practice

3.1 Asking EIU questions

Lists like the one above can look intimidating. How can we set about developing these skills in our pupils? One way to start is by posing appropriate simple questions. It is in the nature of EIU questions that pupils can tackle them at a level appropriate to their own understanding (differentiation by outcome). It is the teacher's role to extend the pupils' understanding, and help to draw out the EIU messages. Here are some EIU questions.

> How much does this cost? Is it worth it? Who paid for it? Where do the materials come from?
>
> What did we do before we had this?
>
> What is the environmental impact of making, using and throwing this away?
>
> How else could we do this?
>
> What are the rules for this?
>
> Who benefits from this? Who loses out?

The following examples illustrate how such questions can suggest useful approaches to specific situations.

Example A: Key Stage 2

Pupils might investigate the cleaning of their school. They could interview the cleaners and caretaker to find out about the cleaning materials and implements. Who uses them? How much do they cost; which is best value for money; where does the money come from? Why have plastic buckets replaced metal ones? Who chooses new cleaners? What hours do they work? How much rubbish is produced, and where does it go?

These questions are not scientific, but they open up possibilities of many scientific activities: e.g. testing different materials, looking at the environmental impact of waste, the design of different implements. Has scientific advance made cleaning easier?

Example B: Key Stage 4

Pupils might find out about the organisation of electricity supply. How does demand vary during the day, from summer to winter? Why is this a problem for supply companies and how do they cope? How can consumers be persuaded by varying charges to alter their pattern of usage? How has the organisation of the industry been affected by privatisation?

The scientific points raised here might include the operation and efficiency of different power stations; the efficient transmission of electrical power; the development of new generating techniques by new suppliers.

3.2 EIU in National Curriculum Science

The Programmes of Study (PoS), together with the examples of activities, give some ideas of where EIU might be developed within science teaching; teachers will also wish to develop their own local examples. Some illustrative examples are given below, each consisting of an extract from the relevant PoS, followed by questions to investigate.

Key Stage 1
"human activity produces local changes in their environment"
What creatures appear in the school grounds when we go home? What plants grow if we leave part of the field un-mown?

"road safety activities"
What protects us from cars? Who controls traffic? Could we have more protection? (A visit from a road safety officer?)

"plants are the ultimate source of all food"
Where do the farmer's crops go? What do the animals eat and where does it come from. How does the cows' milk get to our homes?

Key Stage 2
"the significant features of waste disposal systems"
Who makes compost? Who throws food away? Where do farmers' fertilisers come from? Why are few farms "organic"?

"Properties such as strength, hardness, flexibility...related to everyday uses of materials"
Why does this old car have metal bumpers, this one plastic? How much would you be prepared to pay for a safer car?

"Construct simple circuits"
How can we make a torch for a disabled person? What specialist electrical equipment is already available for disabled people?

Key Stage 3
"the uses of enzymes and microbes"
How are microbes used to make yoghurt? How can we design a new flavour of yoghurt? How could we market this product?

"everyday processes such as corrosion"
How much does rusting cost us? How much does rust prevention cost?

"the use of electronic sound technology...in medical applications"
Who benefits from ultrasonic scanning? Who raised the money for the scanner? Who operates it?

Key Stage 4
"the social, economic and ethical aspects of cloning and selective breeding"
Who makes use of the technology? Who owns it? Who regulates it?

"ionising radiation and its effects"
Who works with radioactive materials? How do they measure and limit their personal exposure?

"how society makes decisions about energy resources"
What types of power station might be constructed locally? What would be their environmental impact? What is the impact of our existing electricity supply system?

3.3 Building confidence – where to start?

How can teachers of science, who may have no experience of industry, help to bring school and industry closer together? How can we develop our own experience, so that our teaching of science leads pupils to a greater critical awareness of the nature of industry and the economic system?

We might start by reviewing our current practice.

- Identify EIU content of current schemes of work.
- Identify further opportunities from PoS.
- Compare with EIU skills checklist.
- Plan to fill the gaps.

We must gradually build up our links with industry. Start in a small way, by identifying people with whom we might work:

- husbands, wives, friends;
- parents, governors;
- careers staff and other colleagues;
- organisations such as SATRO;
- Neighbourhood Engineers, professional institutions.

We need to collect and learn to use new resources:

- from industry;
- other published materials;
- videos, software;
- articles in the local and national press.

We need to develop new activities:

- modify existing activities;
- plan visits, and invite visitors to school;
- devise new practical work;
- encourage work placements;
- set up case studies, projects, mini-enterprises.

The work can be gradually built up, and shared amongst all teachers of science. There is a great stimulus to be derived from working with other colleagues, and with outsiders. Teacher placements in industry, from a few days to a term, can help teachers to build their awareness and understanding. They can be organised independently, through the Teacher Placement Service or your local TEC. In a similar way, people from industry learn a great deal about schools, teachers and pupils when they are invited to participate in school science activities.

4 The Outcome

Science and EIU can enjoy a symbiotic relationship.

EIU can benefit science by providing interesting contexts which show the importance of science; by showing pupils how the science they study is linked to other subjects, such as technology, maths and geography; by encouraging a greater range of activities and teaching and learning styles; and by helping to forge valuable links with industry.

Science can be of benefit to EIU by bringing EIU into the curriculum in a natural way; providing practical activities related to real-life industrial problems; and by encouraging pupils to tackle industrial, social and economic questions in a critical way within the rigorous framework of science. If, after several years of hard work, you can point to success in these areas, you are doing well!

5 EIU: Some Awkward Questions

Q: But I don't know anything about industry, do I?

A: This is true for many, and learning can only be a gradual process – see the section "Building confidence" above. Often, the teacher's role is simply to ask questions, and to help pupils to set off in search of answers. There are not always "right answers", but helping pupils to think clearly and critically about the issues raised will help us to develop our own awareness.

Q: This is all about projects, isn't it? Where is the time to come from?

A: Yes, developing EIU puts demands on teachers' time, and on lesson time. However, EIU does not necessarily imply extended project work. You can inject aspects of EIU into your everyday teaching simply by tagging on questions such as, "How much does this cost?" and "Is it worth it?"

Q: Aren't we in danger of indoctrinating our pupils with an "industry is beautiful" ideology?

A: A quick look at the EIU skills checklist will show that an attempt has been made to make it balanced. We can achieve such balance by inviting into school people with very different points of view – eg a trade unionist and a manager, representatives from British Nuclear Fuels Ltd and Friends of the Earth, or, more frequently perhaps, by using printed materials or audio-visual resources produced from different points of view. Pupils can be challenged to identify differences of economic interest – eg the need to produce goods at a competitive price and the cost of reducing the effect on the environment of producing them – and decide for themselves where they stand on the issues and why. As in science teaching we expect pupils to use rational argument, to think up different interpretations of available evidence and to evaluate them critically.

David Sang teaches at Bishop Luffa High School, Chichester. He has been a member of the editorial teams of SATIS 16-19 and the Nuffield Modular Sciences Project, and written for the Bath Science/Nelson 5-16 and 16-19 projects.

References

NCC (1992) *Science and economic and industrial understanding in Key Stages 3 & 4*, NCC.

Sources and resources

Other NCC publications:

(1990) Curriculum Guidance 4: *Education for economic and industrial understanding.*

(1991) *Teacher placements and the National Curriculum.*

(1991) *Work experience and the school curriculum.*

The ASE publishes SATIS materials (see chapter 20) and their *Annual Meeting Handbook* is a good source of organisations producing suitable material.

The Department of Trade and Industry (DTI) has sponsored the production of two series of 8 videos, *"Electronics Now!"* and *"Innovation: – Wealth from Science & Engineering"*, covering science and technology at KS3 & 4, and post-16. They come free with support materials. Contact: Software Production Enterprise Ltd, 4-7 Great Pulteney St, London WIR 3DF

Neighbourhood Engineers groups link schools with engineers who can help with project work, careers events, visits, etc. Contact: Neighbourhood Engineers, The Engineering Council, 10 Maltravers St, London WC2R 3ER.

The Teacher Placement Service aims to give all teachers the opportunity to experience life in industry at least once in every 10 years. Contact: Teacher Placement Service, Understanding British Industry, Sun Alliance House, New Inn Hall St, Oxford OX1 2QE.

20 Science, Technology and Society

Caroline McGrath

1 Introduction

The field of study known as Science Technology and Society (STS) is relatively new. Yet in a short period of time it has grown into a world-wide movement having significant influence within the teaching of science. Internationally there are many projects based on the STS philosophy and a great variety of acronyms are in use! In the Netherlands we have PLON, in the United States we have CEPUP, and in the Philippines we have STEPS to name but a few, but all have the same common foundation and general approach. Hofstein et al (1988) define STS as *"teaching science content in the authentic context of its technological and social milieu"*. However Solomon (1988) in a review of the structure of STS courses suggests that the generic title of these courses is not very instructive. Commas and conjunctives give no indication of the way in which the three nouns interconnect and interrelate, nor, more importantly, what parts of them are to be taught, and how. Hofstein et al (op cit) suggest that pupils tend to integrate their personal understanding of the natural world (science content) with both the man-made world (technology) and the social world of the pupils day to day experience (society).

These science – technology – society interconnections are suggested by the lines and arrows in figure 1. The unbroken lines represent the connections made in STS teaching materials that provide science content in this integrative way. The design of STS courses therefore requires decisions to be made on how to provide a pattern of material which interprets the broken lines so as to reinforce the unbroken ones.

In interpreting this diagram one should remember that STS material should make science more relevant to the wider world within which pupils live their lives:

- as private individuals, with their own particular needs and interests;
- as members of a local, a national and a global community;
- as future workers.

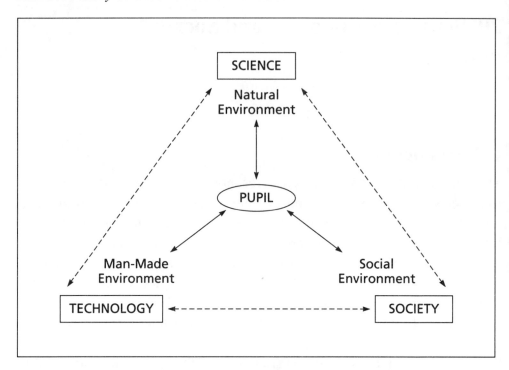

Figure 1 The relationship between Science, Technology and Society and the pupil

Holman (1986) suggests that we are all born scientists, with an innate curiosity about the things around us, the materials that make up the world and about other living things. Some may retain this curiosity all their lives, but others lose it – perhaps because the science they study is too esoteric, too academic, too dry, too impersonal and too far removed from their everyday lives and experience. The introduction of STS material is one way of attempting to stem this loss.

The development of STS in schools in this country can be seen to have happened in two waves. The first wave was aimed at sixth form level and led to the publication, in 1981, of the *"Science in Society"* course and, in 1983, of the *"Science in a Social Context"* (SISCON) course for schools, the latter having its origins in work done in higher education from 1971 onwards. The work of people like John Lewis and Joan Solomon on these projects acted as a springboard, and the ASE decided that it would be a good idea to extend the work on science in society to below 16.

It was as a result of this that the Science and Technology in Society (SATIS) project was set up. What has proved to have been a very important decision was made, that this project would generate not a course but flexible materials. Also, while this first wave was "breaking" the ASE, in its policy statement "Education through Science" (1981), argued that, in planning and developing the curriculum, teachers should show that science can be explored from the viewpoint of its applications, leading to the development of an appreciation and understanding of the ways in which science and technology contribute to the worlds of work, citizenship, leisure and survival. Following this, a working party was formed and given the brief of studying the interactions of science with society in the 11-16 curriculum. The findings were published in an occasional paper called *Rethinking Science*. This report provided many of the themes later adopted by the SATIS project.

In 1983 another project whose concern was particularly to make science more relevant was started, *Salters Chemistry,* based at York University.

The introduction of the GCSE in 1986 was very important for STS, as were similar changes in other parts of the United Kingdom. All GCSE syllabi have to conform to national criteria. In 1988, these criteria said that no less than 15% of the assessment in science should be related to the technological applications of the subject with their social, economic and environmental implications. It was however not permissible to set questions which could be answered without knowledge of the appropriate science. This restriction continues to this day, in the new revised criteria for Key Stage 4 of the National Curriculum, which incorporates the philosophy of STS by stipulating that questions concerning the applications and implications of science must make a significant contribution to each scheme of assessment. Some examples of how the new GCSE syllabuses reflect this requirement, taken from MEG (1994) and ULEAC (1994), are:

> *(candidates enabled to) become confident citizens in a technological world, able to take or develop an informed interest in matters of scientific importance; (MEG 3.2.1 : ULEAC 1.1.)*
>
> *(candidates) recognise the usefulness, and limitations, of scientific methods and appreciate their applicability in other disciplines and in everyday life: (MEG 3.2.2. : ULEAC 1.2)*
>
> *(awareness that) the study and practice of science are co-operative and cumulative activities and are subject to social, economic, technological, ethical and cultural influences and limitations: (MEG 3.5.1 : ULEAC 4.1)*

the applications of science may be both beneficial and detrimental to the individual, the community and the environment: (MEG 3.5.2 : ULEAC 4.1)

the concepts of science are of a developing and sometimes transient nature: (MEG 3.5.3 : ULEAC 4.2.)

Besides being suitable for science courses, many of the resources which have been designed to address STS issues already incorporate much that will now be required by the compulsory five cross-curricular themes of the NCC. Both the Salters courses and the SATIS family of resources could be widely used to deliver these themes.

Unfortunately STS courses at GCSE(M) and AS level have very low entries, but the developments of modular A level and AS level science courses for 16-19 year olds (see p 300) would permit STS modules to be included in courses certificated either as science or as one of the separate sciences.

2 Delivery of Science, Technology and Society

There are two particularly significant features of a science education that recognises the importance of its social, industrial and technological aspects. Firstly, the material taught should be practical and relevant, reflecting the realities of the world. Secondly, the method of teaching should involve pupils in a variety of activities, and demand positive involvement on their part. At present there are a number of projects within the United Kingdom that are based on the STS philosophy and are particularly geared to the teaching of STS :

- The *Salters Chemistry* projects
- The *Salters Science* project
- The *Association for Science Education's Science & Technology in Society (SATIS)* projects.

The Salters projects take as their starting point industrial, social and every-day topics such as Buildings, Drinks, Food or Agriculture. In the case of the Salters Advanced Chemistry course, each unit has a storyline, eg the story of petrol; what it is and how it is made. Science principles are developed from these starting points. The SATIS approach is, in a sense, the inversion of the Salters approach. In this approach, which is more traditional, the industrial and social topics are inserted into a framework of scientific ideas and facts.

3 The SATIS Resources

The SATIS materials are intended to enrich and enhance the teaching of science, and are designed to be used alongside and with existing science programmes; they do not comprise a complete course but are a varied set of resource materials. The units can be used in a flexible and selective manner. Teachers can adapt the resources to meet their own needs.

The following extract, based on work done by Graham Lenton and Andrew Hunt (and many others mentioned in the extract itself), gives a flavour of the development and approach of SATIS material. It describes the development of a unit on DNA fingerprinting for SATIS 16-19 (See page 352) (Hunt and Lenton 1992 pp 22,23).

The unit called 'DNA Fingerprinting' played an important part in the history of the project because it was the first to show that the strategy could produce high-quality, novel resources which clearly met the needs of teachers and students. The authors, Susan Wells and Pauline Lowrie, were biology teachers in Staffordshire and they collaborated with Professor Alec Jeffreys who developed the technique of DNA fingerprinting at Leicester University, Paul Debenham of ICI Cellmark Diagnostics, the company which markets the technique in the UK, and Keith Hadley at the government's Forensic Science Training Unit.

The trial version of the unit was in four parts. The first part explained the science and technology of the technique itself with clear diagrams provided by Cellmark Diagnostics. The explanation was followed by a set of questions. Part Two gave an outline of some of the many applications of DNA fingerprinting, again with questions and opportunities for students to suggest new applications. Part Three included two case studies with DNA fingerprints to interpret. Students had to decide the rights and wrongs of paternity disputes based on the scientific evidence. Finally, Part Four suggested a courtroom drama with students playing all the key roles and in particular trying to explain to the jury why they should or should not convict on the basis of the fingerprint evidence given in the unit.....

One of the most vivid examples of trial feedback for this unit was a video of the courtroom drama played out by the students at a College of Further Education. The video showed that the students had not restricted themselves to the information in the unit when preparing their roles. They supported their arguments for and against the defendant with the help, for example, of reference to recent controversies about DNA fingerprinting in the USA and elsewhere. In this simulated "trial" the defendant was unexpectedly "acquitted" despite the scientific evidence, because of the doubts in the minds of the "jury" created by the well-prepared "defence lawyer."

3.1 SATIS 14–16

The original Science & Technology in Society (SATIS) project was launched in September 1984 with John Holman as Project Director. In June 1986, the first seventy units along with a Teacher's Guide were published. Grouped together in books of ten and known as "decaunits" they quickly found their way into the science curriculum especially as they could be photocopied, copyright waived, in the purchasing institution.

In January 1988, a further thirty units were published. A wide variety of issues was tackled and the names of such units as "Perkin's Mauve", "Recycling Aluminium", "Ashton Island-A problem in renewable energy" and "The Limestone Inquiry" joined the science classroom vocabulary. These units, written by teachers for teachers and validated by experts, introduced many new teaching and learning activities such as role play, case studies, and structured discussion. A further two books (20 units), along with an update on all previous 10 decaunits, linking the units to the National Curriculum, was published in 1991.

3.2 SATIS 16–19

Following the success of the SATIS project the ASE resolved to extend its work to older students. The new project started in September 1987 under the directorship of Andrew Hunt. SATIS 16-19 was similar to the original project in many ways:

- it was funded on a similar basis;
- teachers determined the policy and did the writing;
- the intention was also to produce a varied bank of resource materials rather than a complete course.

There were differences though: the team for SATIS 16-19 had to reflect on the progression from pre-to post-16 courses and discuss the nature of resources appropriate to more mature students who could be following a wide variety of programmes in colleges as well as schools, and who would soon have the responsibilities of being adult members of society.

In many schools and colleges all students are expected to continue some aspects of a broader education through general studies programmes. Various examinations are available as a target for this work, and many courses include a component related to science and technology in society. Thus the SATIS 16-19 project had both to support the general education of all students, and to provide resources to enrich specialist science programmes. The 100 units published in four volumes or files show that the units cluster into themes such as materials, energy, the environment, health, and ethical issues. Older students often have a less prescribed curriculum after the end of compulsory education, with time for private study. They can be expected to use libraries, computer data bases and other sources of information. This allowed the writers of SATIS 16-19 to place more emphasis on guiding the study of students while expecting them to gather the necessary information as a basis for discussion and debate. Resources such as video, slides and audio tape were also developed in conjunction with many of the units. Recently three readers or further units (F-units) have been added to the project, addressing the questions What is Science? What is Technology? How does Society decide? They provide a framework for exploring issues related to science, technology and society, and are extensively cross-referenced to the 100 units in files 1-4.

3.3 SATIS 8–14

This started in 1989 under the direction of John Stringer. There will eventually be three boxed sets of materials from this project. Box 2 was published in April 1992 and is intended for pupils outside this age range. Box 1 will address pupils in the early years of KS2. Box 3 will be aimed at lower secondary pupils in KS3 and will also be appropriate for middle schools, and some top primary pupils.

Box 2 contains ten books, each with five units based on themes, a teacher's guide and a data disc. Each unit is cross-referenced to national curricula. The materials can be photocopied for use in the purchasing institution and the whole project is supported by radio and television programmes.

3.4 SATIS Atlas

This project, organised and edited by Tom Kempton and directed by John Holman, is designed to be used by pupils in the 14-19 age range. The book consists of a collection of 45 thematic maps presented in three outlines; the

UK, Europe and the world. Each map examines a science related topic. The maps are grouped in sections that cover a wide area of the science curriculum. The sections are: Energy, Air Pollution, Water, Weather, Agriculture and Food, Health, and Materials from the Earth. There are questions for students linked to each map.

3.5 Science across Europe

This project was initiated by the ASE in collaboration with BP. The idea is simple. Science is one of the few parts of the school curriculum that is common to all European countries. Each book contains a unit of work with information about a scientific issue of concern to everyone. It invites pupils to collect information, opinions and ideas and exchange them with schools in other countries. A list of participating schools will be supplied each year to schools that register with the project. The first two units are now available – "Acid rain over Europe" and "Using energy at home". Three more are under development – "Renewable Energy", "Drinking water" and "What did you eat?" Each book contains a unit with versions in ten languages, together with maps and data. To date, schools from 20 different countries are involved. The potential of this material is enormous with its emphasis on modern foreign languages and communication.

3.6 SATIS Science with Technology

This project, started in September 1991, is a joint initiative of the Association for Science Education (ASE) and the Design and Technology Association (DATA) with Jim Sage as Project Director. The aim is to develop the close relationship between science and technology and produce curriculum materials to support work in both subjects for pupils in the 14-19 age range. In the past, the T in STS has frequently been subsumed under Science, the first S. Professor David Layton has argued for the recognition of technology as an autonomous co-equal to, and not a subordinate branch of science. (Layton 1988). He suggests that, too often in the past, in much of the western world, technology has been seen as applied science and hence dependent on what is thought to be an abstract, superior commodity. As a result, technology has often suffered from low esteem when compared with science, and science has often been perceived as theoretical and academic. Encouraging the study of science-based examples within the technology curriculum will broaden it and help link it to the science curriculum. In this project the aim is to enhance the status of technology and to increase the appeal of science by making it more practically related to the process of designing and making to meet human needs. The project is trying to put equal emphasis on both the first S and the T in STS and by doing so to address some of the points made in Layton's paper.

The materials will be developed by teachers of both science and technology, working with advisers from industry. The materials themselves will be matched against the requirements of the National Curriculum in science and technology and with other forms of accreditation, including GCSE, BTEC, City and Guilds Diploma of Vocational Education and A/AS levels.

4 The Way Forward

Teaching science in the context of Science, Technology and Society is basically relating science to everyday life. Teachers need to be well-informed about real-life situations relevant to science and technology, use them in their teaching and apply science concepts and principles to explain them. It is hoped that pupils will recognise the many areas of human activity where science and technology are useful, develop a caring attitude for their environment, and appreciate that the world has finite resources and that we all have a responsibility to look after them. For the first time in our history, we have the means to raise the general public understanding of science as well as to broaden the perspective of future scientists and technologists. We cannot afford to miss the opportunity to do so. The way forward is for every science teacher to incorporate the teaching of Science, Technology and Society into the science curriculum and, in doing so, help to create a society that can appreciate the fine balance between science and technology as tools for national survival and the need to manage and preserve the finite resources of the world for future generations.

Caroline McGrath has an Esso fellowship to disseminate and promote SATIS projects and is also ASE Field Officer in the South East of England. She worked in industry before entering teaching and has been a head of science in a comprehensive school.

Further information

A brochure containing the latest information on all SATIS publications can be obtained from the Association for Science Education, College Lane, Hatfield, Herts, AL10, 9AA (tel: 0707 267411).

For further information on the Salters Projects contact the Science Education Group, University of York, Heslington, York YO1 5DD.

References

ASE (1981) *Education through Science*, ASE.

ASE (1984) *Rethinking Science – Teaching Science in its Social Context*, ASE.

Hofstein A, Aikenhead G S, and Riquarts K (1988) *Discussions over STS at the Fourth IOSTE Symposium*, Int. J. Sci. Educ., Vol.10 no 4. p357-366.

Holman J (1986) *Science and Technology in Society. General Guide for Teachers*. ASE

Hunt A and Lenton G (1992) *SATIS 16-19 on trial in the UK*, in, Yager R E (Ed), *The Status of Science-Technology Society – reform efforts around the world*, ICASE Yearbook 1992.

Layton D (1988) *Revaluing the T in STS*, Int J Sci Educ, Vol 10, no 4, pp 367-378.

MEG (1994) *Science (Salters)*, 1994 Double Award Syllabus Code 1774: Single Award Syllabus Code 1775 (General Certificate of Secondary Education National Curriculum – Key Stage 4)

Solomon J (1988) *Science technology and society courses: tools for thinking about social issues*, Int J Sci Educ, Vol 10, no 4

ULEAC (1992) *Science framework:* Common syllabus booklet: General Certificate of Secondary Education – Key Stage 4 1994, ULEAC

21 Science Education and Public Understanding of Science

Robin Millar

1 Why Public Understanding of Science Matters to Science Teachers

Imagine that you are standing in front of a typical science class of, say, 13 year olds, about to begin your lesson. What are you teaching science to the these pupils *for?*

It could be argued that pupils are unlikely, before the age of 16:

a) to have been sufficiently initiated into the activities we call science (concepts and processes) to be able to judge whether or not they wish to continue to devote time and energy to science for the intrinsic rewards and satisfactions it affords;

b) to know enough about themselves and the world in which they live, to know whether or not the things they might want to do in their working or private lives requires an understanding of science up to, or beyond, that afforded by GCSE science.

All pupils should therefore undertake science up to this level so that they may genuinely choose, from an informed point of view, whether or not they want or need to continue with further studies, and, should they choose to, have an appropriate basis from which to proceed. Despite this, however, since less than a quarter of the pupils facing you will continue with any formal study of more advanced science after the age of 16 and even fewer will end up in a career which requires them to make explicit use of science, we quite properly feel that a compulsory science education up to 16 should also be justifiable in its own right. Any such justification must lie in our view that it is valuable, both to the individual and to society, for people to have an adequate understanding of science. The issue of "public understanding of science" – what it might mean and how it might be promoted – is therefore of central importance to all science teachers; for it provides the rationale for most of what they do.

But does the curriculum reflect this? If we look closely at syllabuses and teaching programmes, do they really make us feel that the science education community has a clear view of what public understanding of science involves, and of how we might improve it? Does the National Curriculum provide a suitable science programme? In its origins, such a claim is indeed made. The opening paragraphs of the 1988 Education Reform Act say that the whole curriculum should *"prepare pupils for the opportunities, responsibilities and experiences of adult life"*. But nowhere is the link between this broad aim and the detailed provisions of the Science National Curriculum spelt out. The original report of the Science Working Party did contain a brief section entitled *"The contribution of science to the school curriculum"* (DES, 1988: pp 6-9), but none of the subsequent official documents on, and revisions of, the Science National Curriculum says anything to explain why its particular set of detailed contents (Programmes of Study or Statements of Attainment) is what is required to achieve the general aim above.

I would suggest that it is not at all self-evident that the Science National Curriculum *is* where we would arrive if we took the aim of preparation for adult life seriously – as the *primary* aim of the science curriculum. Instead, it is largely a re-packaging of the science which was previously contained in GCSE syllabuses and, before them, in GCE O-level syllabuses. It has evolved from the kind of science course which was originally designed to prepare a relatively small proportion of pupils to go on to study science at the next higher level. The same is true of recent science curriculum change in many countries. The Australian science educator, Peter Fensham, summarises science curriculum developments around the world since the 1960s in these terms:

> *science curriculum projects ... set out to extend science as it was known in the curriculum of elite secondary schooling to a much wider cross-section of school learners. In other words, the content and topics of these elite science curricula were taken as the knowledge of science that was worth learning more generally, and the projects devoted their energies to devising new presentations and forms of pedagogy which it was hoped would achieve this goal of more and more learners acquiring some of this knowledge.*
> *(Fensham, 1985: p 427)*

Fensham goes on to argue that much more radical rethinking is necessary if "Science for All" is to become anything more than an empty slogan. There is, indeed, plenty of evidence, both formal and informal, that the standard model of science education – a single curriculum for training the future specialist and for promoting public understanding of science – is simply not working. It may be achieving the first goal adequately but it is failing to tackle the second. (Task 1 may help you to reflect on your own teaching schemes in the light of these issues.)

In a recent book which is timely in drawing attention to the bittiness and pointlessness of many school science activities and to the flawed logic which pervades much standard classroom practice, Guy Claxton writes of his *"growing realisation that we do not have a problem with science education; we have a disaster with it"* (1991: vii). He goes on to identify the consequences of this in terms of pupil disaffection, boredom and rejection of science. If we are serious about science for all, about teaching science for public understanding, we need to think harder about what we are trying to achieve: what is the *value* of science education for most pupils as they prepare for life in our society? Why is it worth the effort required to learn? This is not a negative exercise in criticism; only by asking hard questions about *why* we are doing something at all, can we hope to discover *how* we might do it better. The central question I want to raise in this chapter is: how would school science look if we took "better public understanding of science" as its principle aim. What would a science curriculum designed for public understanding of science be like?

2 Public Understanding of Science: a Realistic Goal for Education?

It is worth looking a little more critically at the notion of public understanding of science itself. One very striking characteristic is that it is seen by just about everyone as "a good thing": it is hard even to think what it would mean to be against it. A second characteristic is that there is never enough of it. Articles on it, almost without exception, either argue or imply that the public's understanding of science is less than the authors of the piece would like it to be (see, for example, Durant, Evans and Thomas, 1989; Lucas, 1987, 1988). This is so striking that it raises an inevitable question: is public understanding of science *in principle* unattainable? Is it something which is detectable only by its absence? Or, to put it slightly differently: do we really know what we mean by public understanding of science if we cannot identify any cases of its occurrence?

Task 1 Curriculum Emphases

The Canadian science educator, Douglas Roberts, has proposed a classification of science curricula and textbooks according to their chosen **curriculum emphasis**. He identifies seven distinct curriculum emphases (Roberts, 1988). Different topics within a syllabus, even individual lessons, may have different curriculum emphases; some syllabuses may have one over-riding emphasis. Roberts suggests that one way to identify the curriculum emphasis in a piece of teaching is to consider how you would answer a pupil's question: "Why are we learning this stuff?"

Curriculum emphasis	Outline	Possible answer to: "Why are we learning this stuff?"
Everyday coping	Emphasis on practical usefulness of knowledge, in the home, in everyday situations.	It is practically useful in some everyday life situation.
Structure of science	Emphasis on science as an intellectual enterprise, relating evidence to theory, thinking up models, etc.	It shows you [this course's view of] how science works and what scientific knowledge is like.
Science, technology, decisions	Emphasis on issues involving science and technology; decision-making, social context.	You need to know this to be able to hold a well informed view about an issue which matters.
Scientific skill development	Emphasis on practical skills, scientific processes of thought and scientific method of enquiry.	You are acquiring (practical and/or thinking) skills which you can apply in many situations.
Correct explanations	Emphasis on *outcomes* of scientific enquiry, on reliable knowledge of phenomena.	This is worth knowing because it is true.
Self as explainer	Emphasis on pupils' own thinking, how ideas change, how scientists developed new explanations and ideas.	To help you appreciate science as one way of explaining, and relate this to your own sense-making about the world around.
Solid foundations	Emphasis on current learning as the foundation for fuller understanding.	You need to know this so you can understand what I am going to teach you next lesson/week/year.

- Think about a typical week of your lessons to one class. Which curriculum emphasis (or emphases) underpins the work?

- Is your curriculum emphasis (or balance of curriculum emphases) the same at year 7 and year 11? If you teach at sixth form level, what is the dominant curriculum emphasis there?

- If a lot of your teaching is really (when you are being honest) underpinned by the *solid foundations* emphasis, where does this put the promotion of public understanding of science in your priorities?

Perhaps it is easier to think in terms of individuals than of societies, of *scientific literacy* rather than public understanding of science. This does not solve the problem but it may make it more manageable. We can still ask: what would it mean to claim that a person is, or is not, scientifically literate? What would this person be able to do that a non-scientifically literate person could not? Which groups of people might be regarded as scientifically literate: science graduates; someone with one or more science A-levels; or someone with GCSE Science? If so, does the grade matter? And if a person attains one of these, do they remain scientifically literate thereafter, or does their "literacy" need to be actively maintained?

It is apparent that there are no easy or obvious answers to any of these questions. Scientific literacy is an idealised goal, just like a society in which there is public understanding of science. In thinking about a "science curriculum for public understanding", we need from the outset to set our sights realistically. Even scientists often feel ill equipped to discuss new developments in other branches of science outside their own particular specialisms. They recognise that they simply do not know enough. Nonetheless, the promotion of scientific literacy *can* be a rational and reasonable target for science education, even if it is unattainable. It is the essence of education that we see purpose in travelling hopefully, even though we know that we will never arrive. We can aim to *improve* people's understanding of science, even if we might not want to claim that we can, through science education, make all our pupils scientifically literate, or achieve, in the course of a generation or two, a society in which there is public understanding of science.

We can identify some hallmarks of scientific literacy – ability to use scientific understanding in making practical decisions in everyday life (about such things as health, diet, medication, energy use); ability to understand current issues involving science, as reported in the news media; an understanding of, or at least a feeling for, the major ideas of science, which can shape our views of ourselves and our place in the universe (the geological time scale, evolution, the Big Bang). Or we might prefer not to think of the outcomes of school science as ends in themselves but as a necessary foundation for the later learning of those selected bits of science which different individuals may need for *"specific social purposes"* (Layton, Davey and Jenkins, 1986). In either case, a science curriculum designed for public understanding of science would justify its content and approach in terms of the value of this knowledge and understanding to the learner *in the context of his/her ordinary life*.

3 Aspects of Scientific Literacy

It is useful to separate out a number of aspects or dimensions of public understanding of science:

1 understanding the *contents* of science (the facts, generalisations, concepts, theories, laws, which make up accepted science)
2 understanding about the *methods* of science
3 understanding of *"science" as an institution*

Most people would agree that the first two are an important part of what they mean by scientific literacy. But both dimensions raise further questions: the first about how much science content one must know to be scientifically literate, and what science content it should be: the second of what we mean by the "method(s) of science", something about which there is disagreement and debate. The communication, through the science curriculum, of naive ideas about science method may make a negative rather than a positive contribution to scientific literacy.

The third dimension may require a little clarification. Jon Miller describes it as

> *"an individual's knowledge about what may generally be called organised science – that is, basic science, applied science, and development – and includes both general information about the impact of science on the individual and society and more concrete policy information on specific scientific or technological issues" (1983: p.34-5).*

So it captures that sense of the term "public understanding of science" in which the object of knowledge is science itself, rather than the technical content of science. This might include understanding of what scientists do and the sorts of questions they tackle; of the institutional mechanisms by which findings are reported and become accepted knowledge etc. This sort of "understanding of science" underpins people's ideas about what occupational work counts as science, about who scientists are and what they are like.

Brian Wynne, writing about the public reception of organised science and technology, argues that this third dimension is of crucial importance:

> *... people can never experience science or scientific knowledge as such. They experience it clothed in a variety of concrete institutional interventions and social interests. They quite rationally*

read this tacit "body language" and respond to that. Even if people have a vague general disposition to respect "science" (whatever this means to them), their concrete experience of science clothed in its unavoidable institutional dimensions may cause them reasonably to reject, neglect or suspect it. ..... We can usefully distinguish three aspects of the public understanding of science; understanding its technical content; understanding about its methods (so as to appreciate its limits as well as its powers); and understanding about its forms of institutional embedding and control. The last is arguably as important as the first two for democracy, yet it is curiously neglected in the debate. What scientists often respond to as public misunderstanding of science (in the first sense) may be more authentically seen as public understanding of science (in the last sense), for example when the public appears to reject official scientific assurances about the safety of pesticides, radioactivity, food and water pollutants, etc. (Wynne, 1990: p.28)

Wynne is suggesting that many people do, indeed, understand "science"; and that they see it, quite rationally, as an institution which, when it impinges on their lives, typically does so through inflated claims about technologies and futures, or reassuring pronouncements about safety which are later significantly qualified or even retracted, or dire warnings which are subsequently diluted or quietly allowed to disappear.

What are the curriculum implications of this third dimension of public understanding of science? It seems clear that the issues involved go beyond formal science education, and include science's own presentation of an image, through the activities and pronouncements of scientists and the reporting of science and scientific matters in the media. It is also worth noting that dimension 3 comes close to being a matter of attitude towards "science", rather than understanding of "science". It is not, I think, an appropriate goal of school science education to seek to produce specific attitudes towards "science". We may hope that our pupils arrive at (what we regard as) a balanced and informed attitude towards science, but the focus should remain on the knowledge and understanding which can underpin this. Much of the understanding which is most directly relevant to this third dimension concerns the methods and processes of science. It is to this that we turn in the next section.

4 A Science Curriculum for Public Understanding: Teaching about Science Method

Many of the ideas which have been (and still are) communicated, both implicitly and explicitly, by the science curriculum about the methods of science are naive, and are counter-productive from a public understanding of science point of view. Recent versions of "process science" are a case in point (Screen, 1986; ILEA, 1987). From the APU framework for science performance, and stimulated perhaps by its disappointing findings on application of science knowledge, came the influential, but simplistic, account of science method in "Science 5-16: a Statement of Policy" (DES, 1985) which in turn provided the impetus and endorsement for much of the science curriculum development of the late 1980s. This essentially inductive view of science method, beginning with observation and leading via pattern-seeking to hypothesising, prediction and testing, has been criticised on a number of grounds (Millar and Driver, 1987; Wellington, 1989). Two central objections are that it provides an idealised and misleading account of the ways scientists actually work; and that the "process skills" it identifies as curriculum goals are essentially unteachable (there is no evidence that they are transferable to new contexts and situations, and there are good grounds for suggesting that children can use them from an early age, without instruction).

In the context of this chapter, it is important to emphasise that these are not merely academic quibbles but have direct practical repercussions. People's ability to make sense of scientific information is influenced by their view of science method. An analysis of the media reporting of the aftermath of the Chernobyl incident in 1986 (Millar and Wynne, 1988) reveals the gap between public expectations and expert pronouncements about data and measurement, and demonstrates *"the barrier to public understanding that a widely shared 'naive' view of science constitutes" (p. 395)*. They argue for a less rule-bound and algorithmic view of science method, and for greater emphasis on the "craft" involved in collecting reliable data and interpreting it.

It is, of course, much easier to criticise the teaching of science method than to make detailed and realistic proposals on how to improve it. Perhaps a starting point is to recognise that there are at least two distinct strands within the overall notion of "an understanding of the method(s) of science". One has to do with the collection of data which can serve as evidence in making or supporting a case. This involves an understanding of some procedural concepts, such as accuracy, reliability, repeatability, validity. It also includes the very notion of a measurement itself (the idea of a standard unit and a method of counting), of modelling behaviour in terms of relationships between variables, and of logical reasoning in situations involving several variables. Many of these ideas apply generally to systematic enquiry, not only in the sciences, and centre around the notion of *evidence* and its quality or persuasiveness. The curriculum implications, perhaps, are that practical work needs to give greater emphasis to uncertainty and error. Estimations of accuracy, reliability (the need to repeat measurements) and validity (does it measure what you want to measure) need to become much more commonplace, from an early age. Practical work must avoid any suggestion that there is an infallible method, or algorithm, for gaining the sort of knowledge which can convince other people. This need not involve tasks with a high level of conceptual demand; convincing others that insulator A really is better than insulator B, or that shoe soles X really do grip better than soles Y could, in principle, do the job. I do not, however, believe that useful practical work of this sort can be done in a single 80-minute lesson.

A second separate strand has to do with the role of theorising in science. It involves understanding that the purpose of science is to generate explanations for observations (it is not about making better artefacts, though that may follow from our success in explaining), and that the theories put forward as explanations go beyond the available data. They are conjectures, made on the basis of available evidence and data, but never completely entailed by that evidence. Theories do not *emerge* from the evidence; there is always an element of creative speculation. Understanding this strand of science method involves recognising theory as separate from data, and being able to relate theory and data appropriately (Figure 1). Giere (1991) sees this as central to public understanding and has based an entire tertiary level course around teaching students how to analyze reports of scientific developments using these ideas. Within the secondary school science curriculum, the questions "how do we know?" and "how can we be sure?" need higher profile. At the core of learning about the method of science is the understanding that theory is distinct from evidence: theorising involves imagination and guesswork and risks being wrong. Evidence is used to check, modify and moderate theory.

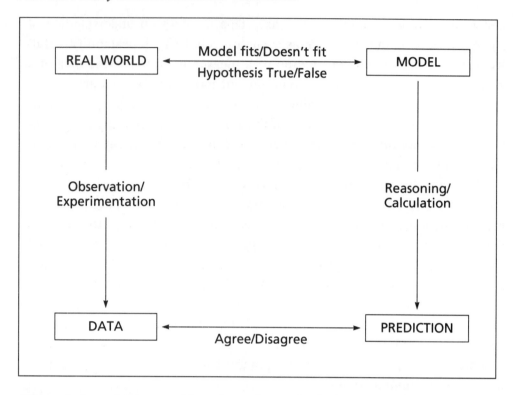

Figure 1 A model for teaching about the method of science and the nature of scientific knowledge (from Giere, 1991)

A science curriculum for public understanding would also provide opportunities to consider some situations where there is controversy – linking ideas about science method back to the third dimension of public understanding. It is important, for example, that learners appreciate that laboratory science deals with simplified situations. Controversies can arise in public science simply because the context has become more complex than the laboratory; they can also arise in core science because data always under-determines theory.

5 A Science Curriculum for Public Understanding: What Science Content?

Let me now turn to the issue of the content of a science curriculum for public understanding. Whilst there is clearly more to scientific literacy than knowing certain bits of science, it must surely be the case that, to be scientifically literate, a person must know some science content. But what content?

One way to begin to answer this is to go back to the question: "why should we be concerned about public understanding of science?" Answers to this question tend to fall into one of three categories (Thomas and Durant, 1987):

1 There is some knowledge which people require because it is useful in coping with aspects of ordinary day to day life. Some scientific knowledge gives you better control over technological artefacts and gadgets and over natural events; instead of simply following blindly a set of operating rules, you have some understanding of what is going on. This could range from knowing how to plan a reasonably healthy diet to simple weather forecasting. This is essentially a *utilitarian* argument for public understanding of science.

2 There is some knowledge which people need to have if they are to be able to take part in democratic decision-making, where science and technology are involved. This might range, at present, from understanding what human activities might affect the carbon dioxide concentration in the atmosphere and hence alter the "greenhouse effect" to holding an informed view on food additives. This is essentially a *democratic* justification for public understanding of science.

3 Science, as much as literature, music, art, is one of the major cultural achievements of our society. We should promote public understanding of science for the same reasons as we support the arts, maintain museums and galleries, and conserve important old buildings. People deserve to be helped to appreciate and understand science and its achievements. This is essentially a *cultural* argument for public understanding of science.

In principle, each of these justifications might be applied to each of the three dimensions of public understanding of science we have discussed above. Here I want to apply it only to the content dimension. Task 2 invites you to look at a list of topics which are found in many current science courses. For each topic, we can ask whether this particular piece of content can be justified as part of a science curriculum for public understanding and, if so, whether its best justification is on utilitarian, democratic or cultural grounds. Perhaps some can be justified on more than one of these grounds; if so, it is useful to ask which is the primary justification. Perhaps some cannot be justified at all. For those which can, it is also useful to consider precisely what it is within the topic that people should know, from a public understanding of science point of view.

Task 2 Science for public understanding: what should be in the curriculum, and why?

Consider each of the "pieces" of scientific knowledge and understanding in the list below. For each, decide which of the following categories you think it should be placed in:

U Everyone ought to understand this at an appropriate level
– for **utilitarian** reasons (i.e. it is practically useful)

D Everyone ought to understand this at an appropriate level
– for **democratic** reasons (i.e. it is necessary knowledge for participation in decision-making)

C Everyone ought to understand this at an appropriate level
– for **cultural** reasons (i.e. it is a necessary component of an appreciation of science as a human enterprise)

X It is **not necessary** that everyone know this. It need not be included in a science curriculum whose aim is public understanding of science.

"Pieces" of scientific knowledge:

	Science topic	Your classification
1	The germ theory of disease	
2	The heliocentric model of the solar system	
3	The carbon cycle	
4	The reactivity series for metals	
5	The electromagnetic spectrum	
6	Radioactivity and ionizing radiation	
7	Newton's laws of motion	
8	Energy: its conservation and dissipation	
9	An understanding of simple series and parallel electric circuits	
10	The theory of plate tectonics	
11	Darwin's theory of evolution	
12	Acids and bases	

As an illustration, consider this example, taken from an article by the Swedish science educator, Bjorn Andersson. Andersson writes:

> *One morning a Swedish radio reporter interviewed people in a line of cars waiting for a traffic light to turn green. He asked them about the signs some environmentalist organisation had set up urging motorists to drive more slowly so as to cause less injury to the forests. One driver said he couldn't see there was any connection between his driving and the forests. After all, the trees were miles away... Some people were enjoying a good dinner together when one of the guests mentioned a relative who lived quite near a big airport. This person had noticed that a bad smell came from the planes at times and was worried about the effect that all the exhaust from the engines might have on her vegetable patch. On phoning the airport authorities she had been told that it wasn't anything to worry about. According to the official the exhaust thinned out in the air and disappeared! ...*
>
> *There is a similarity between these two examples – one system affects another without any visible link between them. ... School science provides some key concepts by which this type of situation can be interpreted, e.g. atom, molecule and chemical reaction. When cars and aeroplanes are running, chemical reactions take place between molecules of air and fuel. Various new molecules are formed, which can, to be sure, spread out over very large areas, and these are indestructible under the conditions prevailing on earth. The new molecules may react with molecules in organisms, sometimes in ways that are harmful.*
>
> *The concepts used here – atom, molecule, chemical reaction should be part of the mental equipment of every pupil by the time he or she leaves school. They are key concepts that help build a rough model of various situations, for example, one's own working environment. These concepts enable us to form a general picture and provide a basis for further enquiry about the details.*
> *(Andersson, 1990: pp.53-4)*

Here Andersson provides a specific justification, in this case, primarily a democratic one, for including some aspects of "the particulate theory of matter" in a public understanding of science curriculum. In the process, he has identified more precisely what he would want to teach: the emphasis should be on using the theory to explain the conservation of matter in physical and chemical change. This general point is worth noting: only if we know *why* we want to include a certain piece of science in the curriculum, can we decide *how* (i.e. with what emphasis and to what depth) we ought to teach it.

Task 2 is a very open-ended one and different readers will reach different decisions and conclusions. Had I chosen different examples in Task 2, the outcomes would also have been different. I would suggest, however, that the following general points emerge from this sort of content analysis:

1 Some familiar science topics are very difficult to justify on any of these three grounds.

2 Very little science content can be justified by a narrow utilitarian argument. Many people live quite successfully in a technological environment with no understanding of science. It is simply untrue that you *need* to understand some science to get by in today's world. Most "really useful" knowledge is *technological,* or even simple rule of thumb.

3 The democratic justification must also be pursued with caution – even scientists feel unable to make valid judgements about areas of science distant from their own. It is, therefore, misleading to pretend that we can give pupils enough understanding to reach truly informed judgements. We can perhaps only aim to provide a *framework* which facilitates the subsequent learning of detail where required. Much more effort is needed to identify what this framework consists of.

4 Many topics are difficult to justify on either utilitarian or democratic grounds, yet we still feel they ought to be part of a general education. People deserve the opportunity to understand these. The justification for including them is largely cultural. We need to give much more thought to how we would approach such topics. If, for example, we were to plan to teach Newton's Laws of Motion *from a cultural perspective,* this might lead to a very different approach from the usual one.

Task 2 begins from the curriculum as it is at present and critically analyses the topics within it. An alternative would be to begin with a clean slate and to look for some general principles for selecting content. This is the approach taken by a working party of the American Association for the Advancement of Science in their report *Science for All Americans* (AAAS, 1989), which forms part of their Project 2061. The AAAS model is structured around a series of large topics, or areas of knowledge: The Physical Setting, The Living Environment, The Human Organism, Human Society,

The Designed World. It includes psychology and sociology, as well as the natural sciences; and it sets out to link science, technology, mathematics and history of science into a single structure. It is a challenging and thought-provoking report, though its proposals may simply be too ambitious to work.

Rather than trying to redesign the whole science curriculum, I want to end this section by exploring briefly the idea of a "framework" of ideas (point 3 above) which can provide a foundation for later learning, whether formal or informal. Andersson has something similar in mind when he writes of *"key concepts that help build a rough model of various situations"*. It may be more productive to think of the science curriculum in terms of "stories" (Andersson's "rough models"), rather than as a collection of "facts" and "rules". The stories are about entities and their behaviours and interactions. Some examples are:

- *a model of matter as particles, to explain what happens in a chemical reaction and why the total mass remains the same;*
- *a model of thermal phenomena as the transfer of energy (heat) from an object at higher temperature to one at lower temperature;*
- *the "source – radiation – receiver" model to explain how one object can influence another at a distance;*
- *the model of transmission of diseases by microscopic organisms transferred from person to person;*
- *the model of changes being caused by a "difference" of some kind*

(Ogborn, 1990).

If models, or stories, like these are to be useful in constructing a curriculum framework, we need also to be able to identify specific pieces of knowledge and experience which are necessary to underpin the story when it is first introduced; and to see how the story can be developed in sophistication or breadth of application as the pupil progresses through the school course. For instance, the particle model used to explain conservation of matter depends on first establishing that gases are matter (they have weight, and so on); it can, if we wish, be extended either towards a kinetic theory, or towards a more detailed account of chemical reactions. The simple source – radiation – receiver model of radiation processes can be extended to an explanation of vision (by adding the idea of diffuse reflection); or to the case of radioactivity where, because the source can be divided up whilst remaining a source, two distinct methods of transfer of effects are possible (irradiation, contamination) with different attendant consequences; or it can be compared with another explanation of action-at-a-distance based on the idea of a field.

A science curriculum for public understanding would aim to help learners to appreciate the *usefulness* of these core models in everyday life – usefulness being broadly conceived, to include the satisfaction of understanding a major idea, and the ability to participate in rational debate about issues of importance, as well as narrow practical utility.

6 Conclusions

The ideas put forward in this chapter are only outline notes towards a science curriculum for public understanding. There is not space to explore the implications of any of them fully. They are intended as a stimulus to discussion. This discussion is, however, urgently needed. The current science curriculum is wasting too much of the time of too many pupils. It teaches most pupils little of lasting value and puts many off science for life. If we want to do more than simply act as a selection mechanism for identifying the minority who can go on to do "serious" science, then we need to contemplate quite radical change.

Suggestions for Further Reading

A recent book which makes a powerful case for radical change in school science, though from a somewhat different perspective from that of this chapter, is:

Guy Claxton (1991) *Educating the enquiring mind: the challenge for school science,* London, Harvester Wheatsheaf.

A very detailed proposal for a science curriculum to promote public understanding of science is provided by the Working Party of the AAAS Project 2061. The initial chapters of their report argue a rationale for science in the curriculum, which leads into a useful discussion of the natures of science, technology and mathematics and of the links between them. Unfortunately the report is published only in the United States, but it is well worth asking your library (or an American friend) to try to get hold of a copy for you. (A second edition of the report was also published by Oxford University Press, but again only in the USA.)

American Association for the Advancement of Science (AAAS) (1989) *Science for All Americans,* Washington, D.C., AAAS.

The AAAS report has much to say about curriculum content. A different approach to public understanding is taken by:

Ronald Giere (1991) *Understanding Scientific Reasoning.* 3rd edition. Fort Worth, Holt Rinehart and Winston.

Here the emphasis is on understanding the role of models (or theories), prediction and testing in science, in order to be able to make sense of reports in the general and specialist media.

Robin Millar taught Physics and science for eight years in comprehensive schools in Edinburgh, before moving in 1982 to the University of York where he is now Senior Lecturer in Educational Studies. His research interests include public understanding of science and the role of practical work in science education

References

American Association for the Advancement of Science (AAAS) 1989 *Science for All Americans,* Washington, D.C., AAAS.

Andersson B (1990) *Pupils' Conceptions of Matter and its Transformations (age 12-16),* Studies in Science Education, vol 18, pp 53-85.

Claxton G (1991) *Educating the Enquiring Mind: the Challenge for School Science,* London, Harvester Wheatsheaf.

Department of Education and Science (DES) (1985) *Science 5-16: a Statement of Policy,* London, HMSO.

Department of Education and Science (DES) (1988), *Science for Ages 5 to 16. Final Report of the Science Working Group.* London, DES and Welsh Office.

Durant J, Evans G and Thomas G (1989) *The public understanding of science,* Nature, vol 340, 6 July, pp 11-14.

Fensham P (1985) *Science for all: a reflective essay,* Journal of Curriculum Studies, vol 17, no 4, pp 415-435.

Giere R N (1991) *Understanding Scientific Reasoning,* 3rd edition, Fort Worth, Holt Rinehart and Winston.

ILEA (1987) *Science in Process,* London, Heinemann.

Layton D, Davey A and Jenkins E (1986) *Science for specific social purposes (SSSP): perspectives on adult scientific literacy,* Studies in Science Education, vol 13, pp 27-52.

Lucas A (1987) *Public Knowledge of Biology,* Journal of Biological Education, vol 21, no 1, pp 41-45.

Lucas A (1988) *Public knowledge of elementary physics,* Physics Education, vol 23, no 1, pp 10-16.

Millar R and Driver R (1987) *Beyond Processes,* Studies in Science Education, vol 14, pp 33-62.

Millar R and Wynne B (1988) *Public understanding of science: from contents to processes,* International Journal of Science Education, vol 10, no 4, pp 388-398.

Miller J (1983) *Scientific Literacy: a Conceptual and Empirical Review,* Daedalus, vol 112, no 2, pp 29-48.

Ogborn J (1990) *Energy, change, difference and danger,* School Science Review, vol 72, no 259, pp 81-85.

Roberts D A (1988) *What counts as Science Education?* in P Fensham (Ed), *Development and Dilemmas in Science Education,* pp 27-54. Lewes, Falmer Press.

Screen P (1986) *The Warwick Process Science Project,* School Science Review, vol 68, no 242, pp 12-16.

Thomas G and Durant J (1987) *Why should we promote the public understanding of science?* Scientific Literacy Papers, no 1, pp 1-14.

Wellington J (ed) (1989) *Skills and Processes in Science Education: a Critical Analysis,* London, Routledge.

Wynne B (1990) *The Blind and the Blissful,* The Guardian, 13 April, p 28.

Glossary

APU Assessment of Performance Unit

ASE Association for Science Education

AT(1,2 etc) Attainment Target (and number) in NC

ATLAS Active Teaching and Learning Approaches in Science

BTEC Business and Technician Education Council

CASE(P) Cognitive Acceleration in Science Education (Project)

CCW Curriculum Council for Wales

CDT Craft Design and Technology

CLEAPSE/CLEAPSS Consortium of Local Education Authorities for the Provision of Science Equipment/Services

CLIS(P) Children's Learning in Science (Project)

COSHH Control of Substances Hazardous to Health

CTC City Technology College

CRE Commission for Racial Equality

DARTS Directed Activities Related to Text

DES Department of Education and Science

DFE Department for Education

DTI Department of Trade and Industry

EIU or (Education for) Economic and Industrial Understanding

EIS Education in Science (ASE journal)

ELB Education and Library Board (in Northern Ireland)

ESG Education Support Grant

GASP Graded Assessment in Science Project

GCE General Certificate of Education

GCSE General Certificate of Secondary Education

GIST Girls into Science and Technology

GMS Grant Maintained Schools

GNVQ General National Vocational Qualification

HIV Human Immuno-deficiency Virus

HMI Her Majesty's Inspector(ate)

HSC Health and Safety Commission

IOB Institute of Biology

IOP Institute of Physics

KS Key Stage (in the NC)

LEA Local Education Authority

LMS Local Management of Schools

LTL Learning through Landscapes Trust

MEG Midlands Examining Group

MISAC Microbiology in Schools Advisory Committee

NC National Curriculum

NCC National Curriculum Council

NSG Non-statutory Guidance (on the NC)

NCVQ National Council for Vocational Qualifications

NVQ National Vocational Qualifications

OHMIC Offices of Her Majesty's Chief Inspector of Schools for England and Wales

PGCE Post Graduate Certificate of Education

PoS Programme of Study (in the NC)

PSE Personal and Social Education

RIDDOR Reporting of Injuries, Diseases and Dangerous Occurrences Regulations 1985

REA Regional Education Authority (in Scotland)

RS Royal Society

SAT Standardised Assessment Task

SATIS Science and Technology in Society

SATRO Science and Technology Regional Organisation

Sc1, 2 etc Science Attainment Target (and number) in NC

SCAA School Curriculum and Assessment Authority (Proposed)

SCISP Schools Council Integrated Science Project

SEAC Schools Examination and Assessment Council

SISCON Science in a Social Context

SoA Statement of Attainment (in the NC)

SS(S)ERC Scottish Schools (Science) Equipment Research Centre

SSCR Secondary Science Curriculum Review

SSR School Science Review (ASE journal)

STS Science, Technology, Society

TGAT Task Group on Assessment and Testing

TEC Technician Education Council

TVE Technical and Vocational Education

ULEAC University of London Examinations and Assessment Council

UFAW Universities Federation for Animal Welfare

Y1, 2 etc Years of education. Y7 is the start of secondary education at 11+

Index